How to Keep Your Classic Tractor Alive

SPENCER YOST

Voyageur Press

For my best friend and loving wife, Rita;
without you this project would have been pointless
and my life would be empty.

First published in 1998 by MBI Publishing Company, 400 First Avenue North, Suite 300, Minneapolis, MN 55401 USA

Voyageur Press titles are also available at discounts in bulk quantity for industrial or sales-promotional use. For details write to Special Sales Manager at MBI Publishing Company, 400 First Avenue North, Suite 300, Minneapolis, MN 55401 USA.

To find out more about our books, join us online at www.voyageurpress.com.

Library of Congress Cataloging-in-Publication Data

Yost, Spencer, 1961–
How to keep your classic tractor alive / by Spencer Yost—1st ed.
        p. cm.
Rev. ed. of: Antique tractor bible / Spencer Yost. 1998.
ISBN 978-0-7603-2951-1 (sb)
1. Antique and classic tractors. 2. Farm tractors—Conservation and restoration.
3. Farm tractors—Purchasing. I. Title.
TL233.25.Y67 2009
631.3'72—dc22

                                    2008023315

ISBN-13: 978-0-7603-2951-1

Editor: Amy Glaser
Designer: Wendy Lutge

Printed in Singapore

On the frontispiece:
Like the perennial Ford versus Chevy war, there are color wars in antique tractors, and they often create some interesting upgrades to antique tractors. Here, a Farmall owner uses his air cleaner's pre-cleaner to share his opinion of John Deere!

On the title page:
Antique tractors are great at helping out on the farm by pulling grain wagons. Here, a John Deere B is bringing in soybeans from the field.

On the back cover 1:
At least one of the brackets for the engine hoist's cable or chain is typically installed under cylinder head bolts, so you'll need to remove any overhead valve components.

On the back cover 2:
Tedding hay is the act of turning the hay to improve drying times. This is a Daros brand tedder, which is a modern and well-respected brand. It is being pulled quite capably by an antique tractor.

On the back cover 3:
Straight, clean design; half moon hitch; and the articulated steering arm are the hallmarks of the Allis-Chalmers model C narrow front tractor. Smaller than the W and D series that followed in later years, its size and capabilities suit it to smaller chores around the farm rather than heavy field work. Today, it is a popular model for hobbyists and restorers.

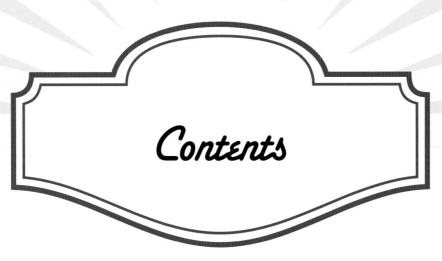

Contents

Acknowledgments

Writing a book is basically a process of learning. Without the teachers in an author's life, the book is impossible at best, a farce at worse. We take what "we" know, which is a distillation of what others know, and pass it on in written form. Therefore, authors of nonfiction are a mixing pot of ideas presented by many people. I am no different and present a book that is a reflection of many lifetimes of learning by many people, all of whom I am indebted to and take the time to thank here.

Special thanks go to:

My children Parker and Elisha for all the times they've lent a hand during "tractor work" and the sacrifice they've made to allow me to write the first edition and for helping with this book.

My mother, Vivian Yost. Her confidence in me is absolute and one of the most consistent things in my life. This is one of the greatest gifts a parent can give and I am grateful to her for it.

The memory of my father, William Herman Yost Jr. He was a better writer than I. I recently had the pleasure of finding and reading many old letters of his and I am more convinced today than I was when I wrote this for the first book: Any literary value you find in this book is because of his teaching, his example, and his genes. I am a mere shadow of him. Thanks, Dad, for reviewing my homework with a judicious and wise use of a red pen.

Steve Sewell has joined me in so many "old iron" trips, shows, and work sessions in our shops that I've lost count. He features prominently in many of the photographs in this book and is a mechanic with few peers. He is a great friend and I look forward to working together for many years.

Brice Adams for being my sounding board and lending a hand with some of the photographs relating to transporting tractors. He has always been a great friend and his sharp mind always forces honesty in my work.

How many times can I say this in my books' acknowledgments without sounding like a broken record? It doesn't matter because it bears repeating. I need especially to thank the men and women of the ATIS antique tractor email list. This book would not have been possible without them. In particular I need to thank George Willer; Francis "Farmer" Robinson; Gene, Larry, and Garry Dotson; and Ed McCullough.

My editor at Voyageur Press, Amy Glaser. Working with an author such as myself who has a full-time job and many other obligations means a lot of missed deadlines and headaches. It is my second book with her and I always felt like I was in capable hands. Thanks for your patience.

The Tri-State Gas Engine and Tractor Association of Portland, Indiana, for so graciously and hospitably hosting our annual gathering and for putting on such a great show. If you can only make one show a year, this is the one you should visit.

The majority of credit goes to all the men and women in this hobby who have been so generous with their time, knowledge, and expertise over the years. It has meant a lot to me and has made my books so much more compelling and useful. Lastly, I want to thank you, the reader. You are always on my mind when I write and my estimations of your expectations motivate and sharpen me. I hope you find this book enriching and enjoyable.

Introduction

When Voyageur Press asked me to upgrade the photographs and captions of this new revised edition of my first book, Antique Tractor Bible, I of course accepted. No one else was going to touch my baby as far as I was concerned and I looked forward to the project. I grabbed my camera gear and started a picture-taking journey throughout the Southeast and Midwest, trying to get the pictures I needed, and began to try to put words to the pictures.

Viewing the change in the hobby and myself, more than a decade later, was harder than I thought. When I originally wrote the manuscript I was fairly new to this hobby, one that was growing by leaps and bounds. It was a hobby populated by older, knowledgeable men and women who knew more than I would ever learn. The attendance projections for antique tractor shows were rosy and most prices for antique tractors were still reasonable. Fuel for our trucks to carry our prized possessions to the show was affordable. My kids were young and interested by anything new and different to them, including antique tractors.

While updating this book I was caught off guard by how much had changed. Fuel is much more expensive, attendance at most antique tractor shows has leveled off, my kids are grown and busy with their own lives, and the prices of many models of antique tractors have risen to the point that many of them are off limits to all but the well-heeled. Unfortunately many of the older, knowledgeable men I've learned so much from have gone on to their greater reward or are simply unable to be active in the hobby.

I wrote the captions and chose the photographs with an eye toward building a bridge to the next era in antique tractor collecting and restoration. This next era will belong to men and women who were born not only well after the tractors were made, but well after these tractors were old and tired. Therefore they look upon them with less nostalgia and more of a historical appreciation. As this newer generation becomes more active, they will bring their own models of antique tractors into the circle of collecting and showing. Models and makes that will be newer and will seem downright modern to us. I view this as good because like all things in life, we merely borrow this hobby from our fathers and mothers only long enough to pass it to our children. Watching them assume their own roles and find their own enjoyment is what will continue to keep this hobby alive and vibrant.

To this end, I tried to update the photos just a little. I included few less steam tractor pictures and tried to include a fewer photos of tractors from the early 1960s. I tried to document skills that we take for granted but younger members of this hobby have never needed or seen. I also tried to write captions a little fuller and richer to help communicate the nuances of a particular topic.

In short, I have tried to make this book more relevant because it is more than a "bible" for rusty pieces of iron. I hope it also serves as a bit of proselytizing for a hobby that I find uniquely American, full of the touchstones for the values we as a nation hold dear. So when you are done with this book, make sure you pass it on to a younger person, planting a seed for the future of antique tractor restoration and collection.

CHAPTER 1

Starting Out

Before the internal combustion engine was invented, there were steam-powered tractors. These behemoths plowed most of the virgin long prairie grass and were in wide use until the early 1920s. These Keck-Gonnerman tractors can be seen most years at the TriState Gas and Engine show in Portland, Indiana.

Like many people, I bought my first antique tractor strictly out of necessity. I had recently acquired some property that included some rough pasture that needed to be mowed. Money being tight, I looked for a less-expensive alternative to a modern compact tractor. An older tractor seemed a pragmatic solution to my problem. While I knew very little about them, I had at least been around a few antique tractors in the past and I knew how capable these relics were. So I consulted advertisements, spoke with folks who knew of tractors available for sale, and attended auctions with absolutely no clue as to what I was doing, what I was going to run across, or how to judge the various machines. I picked out a fine machine, but looking back, I realize that was simply good luck.

After all of this, I became the proud owner of a 1946 Farmall Model A. Popular in these parts (North Carolina), these little tractors were used heavily for cultivating tobacco and mowing grass. As the local rollback company I hired to deliver it unloaded it at the top of my road, I admired its simple, functional form and mulled over the times this machine must have seen. I also wondered, "Why am I the only one who has thought of antique tractors as an alternative to higher-priced modern equipment? Am I a fool or a genius?" I would find over time that not only was I not the only one with this idea, but I was a part of a greater movement discovering the utility and value of antique farm equipment. Owners of antique tractors can apply this utility to diverse activities, such as modern agriculture, hobby or mini farming, estate home maintenance, and antique collecting and restoration.

I climbed on the tractor after it was unloaded. After starting it and beginning down our gravel road, I felt unusual, uneasy, and I couldn't place why. When I acquired this tractor, I was planning on many things happening—the fields getting mowed, for one thing. What I hadn't planned on were all those familiar smells and sounds and the resulting wave of memories they would bring. Most of these machines were built before my time, but their technology was fairly current and common in my childhood. Older, sturdier engines make sounds I haven't heard since they were supplanted with foreign engines and more-modern designs. I drove down the road to my house, my mind adrift in the past.

The next several weeks found me marveling at my new tractor's capability and utility, even though it was nearly fifty years old. I quickly realized this tractor had many thousands of hours left to give with just an occasional repair or two and continued maintenance. What I had found was more than a tractor; it was the fulfillment of many of my needs, most intangible and having nothing

One of the most recognizable and iconic antique tractors is the unstyled John Deere B. Everything about the tractor exudes the qualities that antique tractor collectors admire and look for, such as unstyled tin, spoked wheels with rubber tires, the John Deere name plate, and an easily manageable size. In its day it was ubiquitous and is well remembered by many people.

to do with grass, implements, or maintenance. This venerable machine confirmed my beliefs in the sanity of long-term planning, the true genius of simplicity, and the principle that products should be designed to last forever if you take care of them. It also answered more basic, childish needs to wonder at things that go pop, whir, and bang, to work with logical, mechanical things, and to work with my hands and in the out-of-doors.

Learning how to operate the tractor was easy, mowing was easier, and the grass knew its wild days were over. Finding implements was harder than expected, but possible. Over the months the number of older folks who stopped to tell me about these tractors and how they used them astounded me. I realized how significant these machines were to our country's agricultural and industrial heritage.

Once, while thinking about this, I was overwhelmed by the presence of the ghosts of the men and women who supported our country and fed its children using these machines. There was much more to this tractor than met the eye. There was much to learn, much to appreciate, much to understand.

My journey began.

Frequently Asked Questions

WHAT DEFINES AN ANTIQUE TRACTOR?

Unlike many collectors of antiques, antique tractor collectors and users do not have clubs that give definitive answers as to what vintage, classic, or antique means. The word "antique" is a catchall word for any tractor that seems to be sufficiently old. Some folks reserve the term for the oldest

Antique Tractor Tip

Restoration and You
Restoration means alot of different things to different people, from a quick spray paint and grease to complete disassembly and refinishing. This book uses the words "refurbish," "restore," and "remanufacture" to describe, from least to greatest, the time, energy, expense, and thoroughness involved in a tractor restoration project.

tractors, and use terms such as "classic" or "vintage" for more recent models, but there is no consensus as to exactly when each term applies. Most folks seem to agree that the 1960s models of most makes represent the cusp of antique tractors. Often, models that were produced in the 1960s began their production years in the 1950s and therefore share design similarities with older tractors. Many tractors from the 1960s, however, such as John Deere's New Generation tractors represent a step up in technology and a departure from older designs. While tractors such as the New Generation are covered in this book and are quickly becoming collectible, they are not yet considered antique. These tractors are better described as vintage or classic models.

This book generally uses the term "antique" to describe any tractor designed before the early 1960s, before the influences of the "Space Race" and modern electronics. At about this time, modern metals, alloys, and plastics, as well as automated manufacturing

methods, began to change the appearance and the performance of tractors. Don't despair if your favorite "antique" is a 1968 Ford. This book will still be relevant, valuable, and important reading. It even includes some information on modern tractors, especially as it relates to antique models. If your favorite antique is a 1917 Moline Universal, you will also find this book an important part of your library, though I suggest you think long and hard before putting the tractor to work on chores I suggest and explain later in the book. Tractors this old can be inappropriate and unsafe for many modern tasks.

If I Buy One to Use, Will It Be Safe?

Whether an antique tractor is safe depends primarily on the operator, and on the answer to several questions: Is the operator educated about the tractor and implement? Is the machine in reliable running condition? Is the operator cautious in new circumstances? Is he or she prepared to stop the

tractor if circumstances become uncomfortable? In short, antique tractors are safe, at least most of the time. Yet any tractor can be dangerous and even deadly if the operator fails to make safety his or her number-one priority.

The burden of safety is always on the operator. The operator is the only one who can reach the brakes, ignition switch, or other controls, and is the only person with a full understanding of what the tractor is doing at any point in time. The operator also has the information to decide if the tractor is well suited for the purpose and can make changes if it isn't. Sometimes an operator must decide that the tractor cannot handle the job safely no matter what and must call for other equipment or more experienced operators to be brought in. The operator must ensure that the tractor is in proper running condition and must operate it carefully and appropriately to minimize risks.

Are They Valuable?

Like most antiques, antique tractors have a strong collectible aspect to them, which gives them value. A friend once remarked that you are a tractor collector if you own one that you don't really need. Even folks who live in the city or suburbia with very little chance or no intention of using them other than in a parade or two are collectors. They own antique tractors for sentimental reasons, investment potential, and a raft of other reasons unrelated to agriculture. Since they are collectible, antique tractors generally do not lose value quickly. More commonly, they appreciate, though very slowly.

Antique tractors retain their value as working machines. In addition, individual parts that are much in demand can command premium prices. For example, a complete and working carburetor for many models can fetch over $100. Sheet metal parts are very valuable, as are rear tires in excellent condition. Even at the scrap iron dealer, a middle-sized antique tractor can bring $200–$300. The demand for parts, in addition to whole tractors, helps sustain values. All of this does not mean that antique tractors will show you a tangible profit. Minor appreciation in value is often more than offset by costs for repairs, maintenance, hauling the tractor, and other expenses. Even if you do make a little money, the rate of return would make any good financial planner laugh. You would make a whole lot more money simply investing in most mutual funds. The rare or unusual tractor that makes its owner a handsome profit is more legend than fact, and being lucky enough to own one is a highly unlikely probability. While buying and selling them may not reap many financial rewards, using them in place of modern tractors, in certain circumstances, makes excellent financial sense. Refer to the section on low-capital farming later in this book.

Is More Information Available?

Fortunately, the tractor industry produced tons of documentation, and reams of data are available for

Massey-Harris tractors were painted red and straw yellow in the 1950s, so this Massey-Harris Pony must be sporting the wrong color, or is it? After Massey-Harris and Ferguson merged to form what is known today as Massey Ferguson, some Ponies were shipped to the Ferguson dealers with Ferguson Gray paint, so this tractor is actually painted correctly. Restoring a tractor to its original condition may actually involve some surprises. Understanding the history of your tractor will often help you understand some of the subtle differences and unusual characteristics your tractor may have.

antique tractors. As one would expect, the older the tractor, the more sparse available information becomes. Generally speaking, though, all manner of manuals (parts, operator, and service), sales literature, promotional brochures, and other company documents are available at archives and antique tractor literature dealers. These dealers can be found advertising in related magazines, and many are listed at the end of this book. There are many magazines devoted to antique tractors. Publishers such as this one carry dozens of titles related to these tractors. Agricultural museums, such as Ontario Agricultural Museum, Ontario, Canada, and state libraries, such as the Wisconsin State Library, act as repositories for company archives, including

companies no longer in business. Existing manufacturers, like John Deere, can sometimes give you information about your specific tractor if you have a serial number. John Deere Archives contact information is also available at the end of this book.

The Internet is a convenient source for tractor information, offering data that is easy to search for and store. Another gold mine of information is the men and women who designed, made, sold, or used these tractors when they were new. Often these people are your neighbors and relatives, who are happy to share what they know. Establishing contacts and friends at antique tractor shows and pulls is another way to gain interesting and useful facts. In short, with a little ingenuity and time, the collector,

Like the perennial Ford versus Chevy war, there are color wars in antique tractors, and they often create some interesting upgrades to antique tractors. Here a Farmall owner uses his air cleaner's pre-cleaner to share his opinion of John Deere!

user, and restorer of antique tractors can uncover a wealth of information.

WHY BUY/USE AN ANTIQUE TRACTOR, AS OPPOSED TO A MODERN TRACTOR?

Probably the most common reason people own antique tractors is because they have a need for a tractor but cannot justify the purchase of a new or recent model. Antique tractors are much less expensive to purchase, and later-model antiques are functionally similar to modern tractors. Also, because so many antique tractors were made, parts are still available at reasonably convenient businesses for modest prices. For example, I have always been able to acquire oil filters for any one of my tractors from auto parts stores, which usually had them in stock or could get them the next day. The last reason, and a very compelling one, is the fact that the average handyman can take care of all maintenance and minor repairs himself. Modern tractors, because of solid-state ignitions and computerized diesel fuel delivery, require specialized testing and shop equipment the average person cannot afford.

Often antique tractors fill a niche that modern tractors simply do not fill well. The first example that comes to mind is cultivating. Many antique tractors were designed with this field task in mind and perform it beautifully. Perhaps the finest cultivating tractors ever made were the offset drivetrain antique Farmall tractors like the model A and Cub. Another good example is the Allis-Chalmers Model G. Its rear engine, excellent visibility, and low-slung frame are well suited for cultivating all types of crops that are planted in nearly any kind of row spacing. Antique tractors and their implements also do many other chores, such as plowing, very well and can be an excellent choice over modern tractors for these tasks.

Not all reasons for owning an antique tractor are monetary. Many people who could afford a newer model prefer the look and feel of antique tractors. Others like to tinker with older equipment. Sometimes a tractor is available from a family member or neighbor that carries sentimental value. Some folks have a philosophical inclination toward the reuse and recycling of resources and choose a used tractor for that reason. Many find them a very inexpensive and convenient second or back-up tractor. Others simply enjoy collecting antique tractors as a hobby. The bottom line is that these tractors, while old, usually have lots of life left in them and are not the last resort of desperate, hardscrabble farmers. Folks use them for all kinds of good reasons, reasons that help them prosper, grow, and provide an interesting conversation piece for their garage or barn.

Common Uses and Applications for Antique Tractors

SMALL ACREAGE MAINTENANCE

Undoubtedly one of the primary uses of antique tractors is small acreage maintenance. There are many homes, businesses, and small organizations that own pieces of property too large to maintain with typical lawn and garden equipment. Antique tractors usually fit the bill perfectly in these situations. They are strong, sturdy, and reliable enough to power implements that will perform the tasks quickly and well. Whether you are bush-hogging a small fallow pasture, blowing snow, grading a long precut access road or driveway, or simply pulling a hay wagon for an autumn hayride, virtually any antique tractor can handle the work without strain or compromises in safety.

Antique tractors also work well for livestock maintenance on small farms and stables. Most livestock chores on farms of this size require equipment that is inexpensive, adaptable, and nimble enough to maneuver around barns, feedlots, and bulk storage areas. Because the tractor isn't used constantly in deep tillage, as is the case on crop farms, the power requirements are usually small. Whether you are raising a small dairy herd or keeping horses for pleasure riding, the only task you may ask your tractor to perform that would require significant horsepower is running a hay baler. Even so, with some planning and reasonable compromises, you may be able to find a baler that keeps your PTO horsepower requirements down to 25–35 horsepower depending on terrain, hay type, and condition.

Another proper place for antique tractors is on the pick-your-own or truck farm or for the very large personal garden. Many antique tractors were made specifically for cultivating row crops and vineyards and for orchard work. Antique implements developed for specific models of antique tractors were well suited to these specific tasks and often do a better job than the modern three-point counterparts. Whether the antique tractor is disking an orchard, cultivating bedded plants, or pulling the harvest wagon, you will find it can be a very capable addition to truck farms.

NICHE ROLES AND PERIPHERAL ROLES
Antique tractors can also perform specialty or niche roles: powering grinders, choppers, silo augers and the like, or permanently bearing a specific implement that is difficult to attach or detach. The only limits to the niche roles tractors can fill are safety and your imagination.

In addition to niche roles, an antique tractor can fill a great number of peripheral roles, handling tasks that are too small or awkward for a larger, more powerful tractor, or supporting the main tractor on big jobs when a second operator is available. In this way, there is less need to compromise on size or power with the main tractor. Using an antique tractor for peripheral roles allows you greater freedom and flexibility without tying up a lot of money.

BUSINESS OPPORTUNITIES
Many uses of antique tractors are unrelated or only casually related to agriculture. The local hardware store near me owns a 1960s vintage Oliver tractor and uses it as a forklift. The last Massey-Harris Pacer a friend bought was purchased from a marina that used it to shuttle boats. Antique tractors are also used in the landscaping and nursery businesses, where budgets can be tight because of thin profit margins and seasonal

Many collectors with steel wheel tractors will remove the lugs and install a set of hard rubber parade treads. This allows the operator to use the tractor on asphalt and is generally safer to load on a trailer.

incomes. Some of the most interesting variations of Ferguson tractors I have seen were used at airports to shuttle planes.

COLLECTION AND RESTORATION AS A HOBBY
Owning antique tractors as a hobby is a trend growing very quickly in this country. Every year each show seems to get bigger, and the publications outlining shows get thicker. The collection and restoration of antique tractors supports several hundred shows around the country each year, many Internet Web sites, several

When searching out collectible tractors, remember that high-crop variations, such as this high-crop model A, fetch premium prices. They seem to go up in value slowly but steadily and currently represent a good investment for your money if that is the type of tractor you want. But the antique tractor collecting hobby can be fickle and you'll never know what fad may come along next and knock the fad you invested in off its pedestal. My advice is to invest elsewhere to make money. Tractor collecting should be fun, and if you happen to make money, that's icing on the cake.

dozen periodicals, and scores of new books. The motives for collecting vary, as each collector has his own reasons for owning the tractors. It might be nostalgia, a fascination with tractors in general, an attachment to the past, or perhaps an appreciation of American engineering so much simpler and more rugged than today's. One of the most common reasons is that the tractors evoke fond memories. Perhaps the collector grew up driving a particular model, or is lucky enough to inherit a model that was owned by a highly regarded family member. Whatever the reason, the collection of antique tractors is fulfilling, fun, and less costly than collecting other equipment. An added bonus is that the tractors in a collection can come in handy when the garden needs to be plowed or the road scraped of snow!

THINGS ANTIQUE TRACTORS SHOULDN'T BE USED FOR

Like all things, antique tractors can be misused. Antique tractors were designed to till the earth in preparation of planting and then power or pull the equipment needed to plant, cultivate, and harvest the crop. It wasn't until the post-World War II years that tractor design took nonagricultural uses into account. Even then their application was limited to light-duty, closely related chores such as road grading. When heavy-duty construction and logging equipment were called for, the manufacturers had nonagricultural models available for purchase.

Never mistake the antique farm tractor for a jack of all trades, capable of providing power in any circumstances.

One of the most common—and tragic—applications for antique farm tractors is logging. Tractors are often asked to pull stumps, skid logs, or move large rocks, all things the designers explicitly did not design them to do. To this day, many deaths occur each year from rearward flips caused by an operator skidding logs or trying to pull dead loads such as stumps and rocks with a tractor. In addition, spark-arresting mufflers were generally nonexistent at the time of antique tractors' manufacture, and unless a tractor is retrofitted with one, it presents a fire hazard when used in forests. While logging applications are possible with antique tractors, only very experienced and knowledgeable operators in very controlled circumstances should attempt them. There are many other times antique tractors shouldn't be used—for example, on steep grades. Just be sure to think carefully about any new situation before starting out.

I hope you will find that antique tractors are useful, important, endearing, and a joy to own and operate. Whatever your reason and motivation for owning and reading this book, you will find that it will help you along your own journey in antique tractors. This journey should be one during which you are constantly learning and appreciating these machines for what they can and cannot do. Be safe, have fun, and start your journey today.

CHAPTER 2
A Brief History of the American Farm Tractor

Part of the joy of collecting old farm tractors is the living history that each machine carries. The machine's allure may be that it is the type, model, or actual tractor that your family used, or the tractor simply creates some special resonance within you. Either way, your tractor played a role in changing the face of agriculture and transforming the life of the farmer. Armed with the bit of history presented here, you'll be able to understand a little more about how your machine fits into the big picture.

Before the advent of mechanized farming in the late 1800s, most agriculture was practiced much the same way our forefathers had for thousands of years: Grain was grown, harvested, threshed, and milled with only human, animal, and wind or water power. Fields were plowed and cultivated with devices that changed little over hundreds of years. Land was cleared by hand and often took long years to fully develop. Technology was slow to evolve and was mostly hindered by metal working and smelting techniques that precluded accurate, quick production of parts and implements needed for farming.

Antique tractors come in a bewildering number of makes. Many of them are made under many names by one manufacturer. Cockshutt sold Oliver-made tractors for many years before beginning their own production of tractors, like this model 35, after WWII. Cockshutt also had connections between BlackHawk and Co-op. This model 35 is owned and operated by Ed McCullough.

An often forgotten but long-lived member of the Ford family of tractors is the Fordson. While production was stopped in the U.S. by the late 1920s, production continued in Ireland and England. The tractors were imported into the U.S. until WWII. The imports continued to arrive in the U.S. after the war until the 1960s.

Local metal workers, known as blacksmiths, were the leading edge of metal technology during this time. These blacksmiths, combined with the animal husbands and leathersmiths, formed the backbone of agricultural technology up until the early 1800s. At that time, agriculture was poised to explode into the highly productive and crucial industry it is today. Modern tractors form the main thrust of technological achievement in agriculture and their developmental history is interesting and instructive. In fact, many parallels can be drawn between the modern information revolution we are going through today and the agricultural technology revolution of the late nineteenth and early twentieth centuries.

Threshers and Reapers

One of the most fundamentally important machines designed to mechanize farming chores was the reaper. Reaping is the act of cutting standing crops and laying them in an organized way to be bundled for transport so they can be threshed (which is the act of separating the grain from the inedible parts of the plant). Designs were being generated and prototypes were being built in the late 1700s, though reliable, working, and commercially available models would not be available in the United States until the 1830s. The reaper seems to have appeared in various parts of the world around the same time. Most historians credit American Cyrus Hall McCormick with the invention of

the reaper, but Scotsman Albert Belle probably developed the reaper several years before McCormick. Belle's machine, however, failed to be adopted by farmers in England, where the sociopolitical climate was not ripe for the acceptance of farm labor-saving inventions. Belle's machine was never available on a wide commercial basis.

Obed Hussey, a fellow American, filed a patent for his reaper earlier than McCormick, but McCormick had his design working two years earlier than Hussey. McCormick's reaper, which he advertised through public reaping demonstrations, became a success very quickly. McCormick's strong mechanical and business skills, fortunate location (he was the only reaper manufacturer in the Midwest, the new breadbasket of the country), and blind luck all contributed to his receiving credit as the reaper's inventor.

Other machines to help with the harvesting of grain, such as mowers, winnows, threshers, and binders, evolved at the same time. A combined harvester/thresher was designed around the same time as McCormick's reaper, though very little development of these early machines occurred, and little is known about them. The true combine, invented by Hiram Moore around the mid-1800s, first became prevalent in California, reaching widespread use across the country by the early 1900s. Another change to agricultural technology during the late 1800s was the supplementation of horsepower by steam. Until then, all farm machines were designed to be pulled by a team of horses. Some machines required huge teams that were difficult to handle and expensive to feed and maintain.

Steam Period

Exploration of steam power for agricultural work began as early as the 1850s, and the J. I. Case Company was known to be investigating it as early as 1862. It was not until the 1870s, however, that steam-powered agricultural tools started to become prevalent. Initially, the public was slow to accept steam power in agriculture. Part of the reluctance was a fear of steam engines, created by deadly mishaps in the railroad industry. A related fear was that cinders and coals from the operation of steam equipment would ignite a farmer's harvest. When grain prices dropped around the 1870s, however, farmers sought ways to minimize overhead costs. One of the largest of these costs was the purchase and maintenance of teams of horses.

The first steam engines still had to be pulled to the work site by horses, and they powered equipment such as threshers via a belt. Later, self-propelled equipment became popular. One of the earliest steam-powered agricultural machines, in operation by 1886, was a self-propelled combine, a huge 22-foot model that used two steam engines for power. It was also during the 1880s that self-propelled steam traction engines, or "tractors," became popular. The heyday of steam was from this time until the teens of the twentieth century, when internal-combustion engines

Orphans and oddballs, as the unusual makes and models of tractors are affectionately called, are often some of the most collectible tractors. Silver King tractors are a good example, and this 1937 model brought thousands at the Don King auction where this photo was taken.

became the power source of choice for agricultural tractor designers. While internal-combustion engines would not match the huge horsepower outputs of the larger models of steam traction engines for many years, steam's decline was swift. Internal-combustion engines were lighter, cheaper, safer, and more maneuverable. Steam power was in decline by 1920, though units continued to be made and sold for quite a while longer.

Internal-Combustion Engines

Internal-combustion engines can trace their development as far back as the 1600s, when gunpowder was experimented with as a power source. Various inventors toyed with internal-combustion engine design through the 1700s and into the 1800s. These designs produced one-off engines used for specific applications, and their lack of efficiency and effectiveness was such that no particular design became popular or widely used. Practical, economical, and reliable designs were not developed until the 1850s. Even then, production methods and reliability in engineering were not in place until N. A. Otto, one of the most significant early developers of internal-combustion engines (and from whose name we get the word "automobile"), designed his four-stroke engine in the 1860s. By 1871 he was selling several hundred units a year. By 1890 his patents had expired, and by 1900 more than a hundred companies were manufacturing internal-combustion engines. Fuels for these very early engines varied by designers, as some ran on gas from coal, some ran on other distillates from oil, and some even

The Minneapolis-Moline model R was introduced prior to WWII and is one of its most popular models with well over 30,000 produced. Francis Robinson of Shelbyville, Indiana, owns this particular tractor.

ran on turpentine. The discovery of a cheap, readily available source of petroleum distillates occurred around the same time, when oil wells were successfully drilled in Pennsylvania in 1859. These discoveries helped remove fuel constraints from designers such as Otto and also helped to motivate further development in internal-combustion engines.

Steam tractor manufacturers experimented with internal-combustion engines earlier in these engines' history. Case was involved, as were others. Generally, however, John Froelich is credited with building the first successful tractor powered by an internal-combustion engine. Froelich would become the founder of the Waterloo Boy Gas Tractor Company, the company John Deere would purchase in 1918 to enter the tractor market.

Steam-powered tractors and internal-combustion powered tractors first competed against each other in 1908 at the Winnipeg, Manitoba, field trials. Every year more internal-combustion engine tractors would enter the competition. By 1912, it was clear to all in attendance that the future of tractors involved only those powered by internal-combustion engines. By 1920 fewer than 2,000 steam-powered tractors were being manufactured annually.

Diesel Engines

Although it wasn't until the 1950s and 1960s that diesel engines became popular as powerplants for tractors, the diesel engine was invented early in internal-combustion engine history. German engineer Rudolph Diesel

designed a working compression-ignition engine in the 1890s and by 1900 had several licensees across the world manufacturing his design. In America, Adolph Busch, the brewer, licensed Diesel's design for $1 million and became America's only diesel manufacturer. Shortly thereafter a second American, George Dow, was licensed to manufacture units. Though he eventually stopped making diesel engines in 1915, his engineer, Art Rosen, started working for the new company, Caterpillar, formed by the merger of Holt and Best. Caterpillar would spend years and over $1 million finalizing its designs for a diesel engine crawler. By 1932 Caterpillar had a successful diesel engine product. Some problems with the fuels and oils used in the engine threatened to ruin Caterpillar, but problems were sorted out through cooperation between Caterpillar and Standard Oil. Within a few years every tractor manufacturer had a diesel engine, and by 1960 more than half of all tractors used diesel power. Today, new gasoline-engine tractors are nearly impossible to purchase.

Advent of the Modern Tractor Industry

THE 1920S
The biggest sensation in the early days of the tractor industry was the Fordson Tractor, introduced by Henry Ford in 1917. This tractor did for tractors what his Model T had done for the automobile industry. While certainly not the best, biggest, or most advanced

tractor, it did have two excellent attributes: It was fit for the purpose of farming and it was affordable. Until this time tractors were quite large and expensive, and owners of smaller farms often could not justify the purchase. Ford with his Fordson brought tractors to the small farm. He owned the tractor market well into the 1920s, when International Harvester, through its McCormick-Deering division, released its Farmall (later known as the "Regular") model in 1924. While the McCormick-Deering Model 10-20 had been giving the Fordson some good competition in the marketplace, the Regular cinched its fate. Production of the Fordson ended domestically in 1928. Henry Ford indicated it was because he needed production resources for his new Model A road car, but the fact of the matter was the Fordson had run its course on American farms.

THE 1930S

From the close of the 1920s until the beginning of World War II, the Great Depression brought challenging times to all tractor manufacturers. This was also a time of experimentation. It was during this time period that Allis-Chalmers began its work with pneumatic tires. John Deere and IH were beginning to design and produce their new letter series tractors, which were to become highly successful for both manufacturers. Caterpillar produced the first diesel tractor in 1934, and other advances in metallurgy and machine work in the 1930s led to tractor engines

The older the tractor, the simpler they were. This early 1930s John Deere D had three speeds: high, low, and reverse.

with greater reliability. Many changes on the financial side of the tractor industry occurred as well. Because of the poor business environment, most tractor manufacturers were teetering on the brink of ruin. Many of the most financially distressed companies were bought and merged into the bigger ones that were weathering the storm. R. B. Gray, in his book *The Agricultural Tractor: 1855–1950* claims that in 1933 the number of principal manufacturers was nine, down from ninety in 1920.

The difficult times of the early 1930s created strong tractor manufacturing companies, which then began a prosperous time, beginning in 1935. As more farmers began seeing a turnaround in the Depression, they began committing to the replacement of the last of their work horses. Smaller models, such as the Allis-Chalmers B, began selling well as

the smallest farmer, neglected by tractor manufacturers in the past, began mechanizing his farm. In retrospect, tractors designed in the 1930s resemble modern tractors in style, functionality, and auxiliary systems. Row crop designs, modern in-head valve engines, hydraulics, three-point hitches, and styled sheet metal all began appearing during this time. Steel wheels were quickly becoming obsolete and horsepower-to-weight ratios began to climb. By the beginning of World War II, tractors barely resembled their predecessors built a scant 15 years earlier.

Antique Tractor Tip

Local Research
Use your local library as a source of historical information relating to your tractor and local agriculture.

THE 1940S

The specter of war brought additional hardships and unique opportunities at the turn of the decade. The hardships included raw material and labor shortages, but tractor manufacturers also saw a new market in the defense industry. Industrial and construction variants of agricultural tractors made excellent equipment tugs and construction vehicles for the armed forces. The defense department generated a great wave of advances in manufacturing and machine design. All of this set the stage for one of the most prosperous tractor manufacturing episodes in modern history: the post-World War II years.

GIs returning from the war, a lack of new tractors during the war years, and a strong economy after the war created a tremendous demand for tractors. Manufacturers were experiencing a heyday that had never been seen before, and it didn't go unnoticed by investors and venture capitalists. Many new manufacturers of tractors incorporated during the post-World War II years. The excessive demand, advances in manufacturing created by the war, and new ideas about modular design and assembly made getting into the tractor industry relatively easy for anyone with money and a little experience. While the larger manufacturers, such as IH and John Deere, simply made improvements to the existing model lines or added a model or two, some manufacturers were revamping their line and making significant changes. Oliver, for example, began making diesel engines widely available on most of its tractors, and Massey-Harris expanded its line to include the most powerful production tractor of its time, the Massey-Harris 55.

THE 1950S

The 1950s can be categorized as a period of tremendous change for tractors, primarily in size, type of fuel, and horsepower rating. John Deere and IH models rapidly changed and increased in both size and horsepower. Both companies began making diesel engines widely available. Companies such as Allis-Chalmers were gaining market share as they revamped their lines. John Deere, while still using the two-cylinder design, was making wider use of its vertical two cylinders, improving upon the three-point hitch and, by the end of the decade, designing its first tractor line that would depart from the two-cylinder legacy. Power-boosting equipment such as turbochargers became more common. The use of distillate fuel had nearly disappeared by the end of the decade and diesel fuel continued its slow but steady rise to prominence as a tractor fuel. In 1959, the specifications for the three-point hitch became standardized.

Important during this period was the rapid increase in the number of models built using off-the-shelf components and third-party systems. For example, many manufacturers were no longer relying on their own engines, but were outsourcing engine production to established companies such as Buda, Continental, and Waukesha. Steering components were often made by Ross, air cleaners were made by Donaldson, transmissions may have been outsourced to Fuller or Chrysler, hydraulics to Dowty, and wheels and rims to F&H. The list goes on, as nearly every tractor system has a counterpart available on the market, ready-made. This trend continued and was part of a growing trend in American business to outsource production of parts and systems that were not critical to the company or the quality of the product. The biggest makes were among the last to outsource, but by the 1970s even John Deere and IH were doing it. In the 1970s, IH was even buying whole tractors from Steiger and simply rebadging the tractor as its own. This practice helped standardize tractor design and keep smaller tractor manufacturers in business.

THE 1960S

While the tractors made during the 1960s are beginning to stretch the definition of antique, many important points need to be brought up here. This decade saw great advances in tractor safety. ROPS, or roll-over protection structures, became widely available during this period. Many systems and designs that we associate with modern tractors were developed and incorporated into product lines in the 1960s. Hydrostatic transmissions, four-wheel drive, and articulated tractors all came into their own in the 1960s. The use of cabs and dual rear wheels became more commonplace. The first Japanese tractor, the Kubota, entered the U.S. market during this decade. In retrospect, the 1960s represent a decade of fundamental change in tractors. For this reason, you will find that many tractor models from the beginning of the decade are considered antiques, while those models introduced at the end of the decade are considered modern tractors. Of course, years from now the tractors made today will be considered antiques, but currently the late 1960s tractors are more representative of today's tractors than they are of antique tractors.

Major Manufacturers and Recommended Models

The remaining section of this chapter will discuss the histories of the individual manufacturers of tractors and will then briefly

Certain styles and parts on an antique tractor immediately give away the age of the tractor. The iron steering wheel, the cast-iron seat, and the lack of sheet metal styling should immediately tell you this tractor is no younger than the mid- to late 1930s.

mention some notable models for each line. Please note models that make good work and production tractors are mentioned first, followed by "cross-purpose" tractors. These tractors are good work and chore tractors, but they also have a collectible aspect to them. The tractors best suited for collections are mentioned last. For example, I mention the John Deere A in the cross-purpose paragraph of John Deere's section. That is not because the A makes a terrible work tractor and should not be considered as such, nor do I think the John Deere A is not a collectible tractor. I include it in the cross-purpose paragraph because it can be both at the same time. If your favorite model isn't included where you think it should be, don't despair. Simply realize that your favorite model

fits well for the section I mention it in. That doesn't preclude it from being able to fulfill any number of roles.

JOHN DEERE

Deere and Company has a long history, beginning with John Deere himself. He perfected the steel self-scouring plow. With his partner, Leonard Andrus, Deere began selling plows in 1837, and the next ten years saw tremendous growth in their business. By 1847, though, Deere had decided to leave the partnership. He then formed another plow company, which also did well. He sold this company in 1858 to his son-in-law and son, but stayed close to the company. The company was reformed again as Deere and Company in 1864, with John, his son, Charles, and his daughter, Emma, and her husband as principal

John Deere's three-point hitch system evolved throughout the 1950s. This hitch is one of the latest variations before the introduction of the New Generation tractors in the very late 1950s. The center link is missing from this tractor because original center links are scarce and are unfortunately often stolen off the tractors at shows. Otherwise, this photo shows a very complete hitch system. The hitch, live PTO, and hydraulics featured here herald the beginning of modern large tractors that have changed very little since then.

shareholders. This company continued to grow and prosper, and by the end of the century it was the largest plow maker. The company also carried a full line of other farm implements, some of which were made by John Deere, some acquired through merger and partnership, and some bought from other companies and simply resold. All of this set the stage for their movement toward tractors.

In the mid-1910s, John Deere began experiments into the feasibility of an inexpensive tractor. The principal designer was Joseph Dain, who spent approximately two years studying, building, and redesigning prototypes of a three-wheeled, all-wheel-drive tractor that would be John Deere's first tractor. Although it was more expensive than hoped, 100 units were built and shipped to New York. While these units were being built, Deere purchased the Waterloo Boy Tractor Company. This gave John Deere a ready source of affordable tractors that were already proven and accepted by the market. Through the Waterloo acquisition, the talent and experience and design knowledge was now on staff to begin the prototype of the first truly successful John Deere tractor: the model D. By the end of the 1920s, the model GP was on the market, and by the mid-1930s, the first models of the post-Depression letter series, for example, the A and B, were on the market. Deere and Company was well on its way to becoming the legend it is today in agricultural equipment.

Models to Consider Purchasing

For a good working tractor, the New Generation John Deere models are best. The model 2010, and from the 20 series, the 2020 and 4020, enjoy particularly favorable reviews. These were made from approximately 1961 through the late 1960s and have the three-point hitch, live hydraulics, and PTO. Many are still in service today on American farms. Another popular working tractor is the 435. It shares the drivetrain and features of John Deere's 430 series tractors, but is powered by a Detroit Diesel engine.

Several John Deere models also serve the role of cross-purpose machines—that is, they can either be held as collectible antiques, or the owner can use them for typical farm and estate chores. The younger letter series tractors from John Deere (models A, B, M, and so on), are excellent cross-purpose machines. Those made in the later years have hydraulics (Power-Trol or Quick-tach), which helps them in the field a good deal. The models in the older letter series have the older style hydraulics, which were simply up-and-down hydraulics, which means you could raise the implement or lower it, but you couldn't use the hydraulics to adjust the implements between these two extremes.

John Deere tractors have a fanatical following in collecting circles, and many tractors that graced the product line of yesteryear are now highly sought after and command premium prices. As far as an entire line of models is concerned, the 30 series (the 730 particularly) has strong collectible value. Any unusual configuration of a more common model or line is likewise valuable. Collectors should look into earlier unstyled letter series, the model GP, and the first tractor made by John Deere, the Waterloo Boy. One of the most valuable tractors in the collecting industry is one of the Joseph Dain-designed and -built three-wheeled tractors. It is believed that only one of these exists. Less rare, but still plenty valuable, are the experimental models such as the precursor to the GP, the model C. "Spoker" model Ds (steel-wheeled model Ds from the earliest years of production) and any John Deere tractors with very low serial numbers tend to be highly collectible as well.

INTERNATIONAL HARVESTER

The manufacturer of one of the longest and proudest lines of antique tractors of all time, International Harvester has enjoyed a reputation of making excellent tractors while maintaining competitive pricing, having complete implement side lines, and, after the shakeout of the antitrust suit, having reputable dealer networks. While the company sold its tractors under a few different brand names, such as McCormick-Deering, Farmall, and International, all brands were manufactured by the same company and shared the same design and manufacturing qualities. IH began its history as several dozen smaller companies that existed around the turn of the century. In one of the biggest business mergers of the early 1900s, McCormick, Deering, Plano, and others merged into International Harvester. It is estimated that right after the merger, 85 percent of the nation's threshing and reaping manufacturing capability existed under the IH umbrella. The Sherman Antitrust Act was written in part because of the IH conglomeration, and after the antitrust suit the company was forced to divest parts of the business and consolidate dealerships. These changes were small compared to the size of the company and IH was left unscathed by them.

In 1939, IH had a huge year: It released its Letter Series tractors. These were IH's first "modern" tractors and their first to sport the new "styled" sheet metal that John Deere popularized on its "styled" letter series tractors. All the tractors of the Letter Series became very popular and quickly claimed the lion's share of the tractor market. Beginning in 1947, enhancements added to the Letter Series tractors caused a product name change by adding the word Super, and the tractors became known as the "Super-M," "Super-A," and so on.

In the mid-1950s IH retired the Letter Series, and replaced it with the Hundred Series. The model A became the 100, the Model C became the 200, H the 300, and so on. These tractors continued to evolve into the mid-1960s. The only drawback to IH tractors was the lack of the three-point hitching system. IH engineers, confident of their two-point Fast Hitch design, were resistant to the three-point design. IH was, in fact, the last major manufacturer to incorporate the three-point into its tractors, though it was eventually blended into all the product lines by the mid-1960s. It should be remembered that IH also made very successful industrial variants of its farm tractors and had complete construction and crawler lines as well. These machines often shared engines and drivetrains with the farm tractors.

Models to Consider Purchasing

If you are in the market for an IH work tractor, your best bets are the Hundred Series (that is, the 100, 200, 300) and later tractors. These tractors did not come from the factory with three-point hitching systems, but aftermarket three-point hitch kits are available for them. These tractors already had live PTO and hydraulics and had the Fast Hitch that IH designers favored. In fact, if you prefer to keep the Fast Hitch on your International tractor, some Fast Hitch implements are still being designed and made and are available new from some of the smaller, independent implement makers.

Any of the IH Letter Series Farmall tractors, or any of the corresponding standard-tread W or WD Series McCormick-Deering tractors, make a good compromise between collectibility and workability. In particular,

Antique Tractor Tip

Don't Forget Your Dealer
Some farm equipment dealers were very conscientious about hanging on to old manufacturer's literature and brochures. In return for your business, the dealer may be willing to allow you access to the archives and collections for your research and study.

Coming up with a wheel hub that maintained adjustment while providing a firm seat to the axle was a design challenge for all companies. This John Deere G shows Deere's taper lock design that locks more firmly as the unit is tightened.

you should pay attention to the post-World War II Super Series Farmall tractors. Some Letter Series Farmalls had hydraulics as an option from the start, such as the model M; others didn't get hydraulics until the Super version came out. Farmall's drive-reduction system, the Torque Amplifier, first became available on the Super M, which became the first tractor to have a live PTO. If the intended chores for your tractor require a lot of hydraulically operated implements, you should focus on the post-World War II Super Series tractors. If your chore list includes a lot of belt and drawbar types of work, then any of the Letter Series Farmalls or W Series McCormick-Deering Tractors should work well.

If you're looking for collectible IH tractors, you have a lot to choose from. A list of the early tractors of the IH line reads like a who's who of antique tractors. Older tractors, such as the Titan, Mogul, and Regular, were important tractors in the development of gas-powered traction engines. The F Series tractors are also collectible and hold a mystique for many collectors. Of special note are the unusual F Series, such as the F-30 Cane Tractor and any of the high crop styles of F Series. The F Series models also saw the change in color from gray to the now well-known red of International Harvester. IH was the first with a rear-mount PTO, and many of these early models have them.

MASSEY-HARRIS, FERGUSON, AND MASSEY-FERGUSON

Massey-Harris, the most prominent Canadian tractor manufacturer, is one of the two main predecessors of the modern Massey-Ferguson brand.

Massey-Harris began as the conglomeration of two popular Canadian implement makers, Massey and Harris. The pivotal point in Massey-Harris history that put in the tractor business was a buyout of the J. I. Case Plow Works. This gave Massey-Harris a foothold in the American market. But first, I must briefly mention the J. I. Case Plow Works and its relationship to Massey-Harris.

The J. I. Case Plow Works was partially owned, heavily influenced, and at one time led by the Wallis family. The Wallis family had married into the Case family, and the plow works were once part of the Case Threshing Machine Company (a full discussion of this falls under the Case sections later). Important to this discussion of Massey-Harris is the fact that the Wallis family led development efforts that produced two of the most forward-thinking tractor models of the time: the Wallis Bear and the Wallis Cub. These tractors, especially the Bear, were never big sellers, however, and the J. I. Case company continued to struggle through the 1920s. The J. I. Case Plow Works' fortunes, and the health of the company, declined until 1928, when a deal was inked with Massey-Harris to merge the two companies. The resulting company retained the Massey-Harris name. Now Massey-Harris had a viable tractor line, manufacturing tooling, and space for tractor production, and some forward-thinking designs and staff to assist them

While the Bear and Cub were no longer sold, Massey-Harris

adopted many of their attributes, including the "channelized" frame created by Wallis. This can be found on many of the Massey models prior to World War II. This design used a single cast (though Wallis first used boilerplate) frame member that served as a crankcase and frame. Massey continued to develop and build solid, reliable tractors that became, with Harry Ferguson's influence, the Massey-Ferguson tractors of today.

Harry Ferguson was primarily an inventor, and he created probably one of the most innovative designs in the tractor industry: the draft-sensing three-point implement hitch. This hitch allowed the hydraulic system to react to ground conditions during tillage. This was a breakthrough in tractor design, but it was protected by patent, and it was 30 years before this hitch was widely available on most models of tractors. Ferguson first used the hitch in Europe in tractors of his own design, but it was the North American market where the hitch found its home.

In the mid-1930s, a handshake agreement with Henry Ford put the hitch on all of Ford's new tractors, including the 9N he had just begun designing. This deal also allowed Ferguson to start marketing Ford's tractors through his own dealer network. In 1947, the handshake deal collapsed, and Ferguson started making his own tractors, the TO and TE models 20, 30, and later, 35. This continued until 1957, when all of Ferguson's tractors and Massey-Harris's tractors were rolled together into one line, the Massey-Ferguson line. This

brought together some of the most innovative people in the tractor industry: those in the Massey-Harris line, with its breadth, width, and original Wallis influences, and Harry Ferguson, arguably one of the most influential men in tractor history.

Models to Consider Purchasing

While the product line of Massey-Harris and Ferguson was not quite as neat, clean, and organized as IH or John Deere, there is nonetheless some rhyme and reason to it. Models that make excellent work tractors are any of the postmerger tractors that carry the name Massey-Ferguson. To this day, very little distinguishes modern small utility tractors from the Massey-Ferguson 35s and 65s of the late 1950s and early 1960s. These two models and the later 135s and 165s are tremendously useful tractors. Out of all the pre-1970 tractors available for use as work tractors, I would have to choose the model 35 and model 65 as the best all-around small and mid-sized work tractors. You simply can't go wrong with one of these unless your power requirements are above 75 horsepower.

Among premerger makes, you should consider any model of TO Ferguson tractor for a work tractor, particularly the TO-35, which is the predecessor to the MF 35. The triple-digit Massey-Harris tractors like the Massey-Harris 333 and 444 offer the three-point hitching system, live hydraulics and PTO, and excellent horsepower-to-weight ratios, and they are excellent work tractors. The older Masseys make

John Deere's Powr-Trol system evolved over the years. This system on a model G shows the 1940s/early 1950s incarnation.

good belt tractors, particularly the Challenger, Pacemaker, and the 101 Juniors and Seniors. Any of the Massey-Harris models would serve well as drawbar tractors, but particularly good are the 44 and 55, which were made in appreciable numbers in the wheatland style.

Almost any of the tractors mentioned above would also serve the dual role of collector piece and work tractor. In addition, one could consider the Massey double-digit tractors, the 22, 33, 44, and 55. These tractors are collectibles in their own right, but many came with the three-point lift late in their production years and had live hydraulics. One other tractor to consider in this category is the model 11, or Pony Model tractor. While very lightweight and underpowered for most farm tasks, this tractor nonetheless

Cooperation and rebadging of tractors began in earnest in the 1950s. In the late 1950s and early 1960s, Massey-Ferguson sold this model 95 built by Minneapolis-Moline. The identical models in the Minneapolis-Moline lineup were the G-VI and G-704. These were available as a two- or four-wheel drive, and the fuel option included LP, gasoline, and diesel.

adequately performs many light-duty chores around farms and estates. In addition this little tractor is easy to transport and is considered an excellent addition to any collection.

From a purely collectible point of view, the pre-World War II Massey models have excellent collectibility characteristics. Of particular note is Massey's foray into the field of four-wheel drive with the model GP. While underpowered and ahead of its time, it is now a highly collectible tractor. The Pacemaker and Challenger make excellent collector pieces too. Of course, any Wallis tractor is highly prized as a collector tractor. The Equine line of Massey-Harris tractors from the late 1940s and early 1950s are highly collectible. These are

models 16, 21, and 23, known as the Pacer, Colt, and Mustang, respectively. The model 11 Pony mentioned earlier is part of this line, but is not quite as valuable because it was made in greater numbers. The Pacer, Colt, and Mustang all had total production numbers lower than 4,000. Finally, if the collector is looking for a large tractor, the Massey-Ferguson 95, 97, and 98 were made in limited quantities, and the fact that they were actually made by Minneapolis-Moline adds to their mystique and collectibility.

J. I. CASE THRESHING MACHINE COMPANY

J. I. Case Threshing Machine Company had a long history of agricultural technology leadership

by the time gas tractors were becoming feasible. Case was a leader in the threshing machine business and made early forays into steam power, with both horse-drawn and self-propelled machines. Case entered the gas engine tractor market in 1911 with the model 30-60, though as early as 1892 an experimental model existed and was abandoned after the powerplant, an Otto four-cycle engine, performed poorly. Case models continued to improve and evolve from the 30-60 until the late 1920s, when Case abandoned the cross-mount engine design—in which engines are mounted perpendicular to the primary axis of the tractor's frame—with the introduction of the Model L.

Related to the threshing machine company was J. I. Case Plow Works, which Case began with a man by the name of Ebeneezer Whiting in 1876. This company, and the main threshing company works (both bearing the Case name) lived amicably side by side until the early teens of the twentieth century. At that point, the threshing machine company began manufacturing and selling deep tillage equipment under the Case name. This was done because tractor buyers wanted an integral line of tillage equipment for the new tractors they were buying. Up until then, the plow works covered this market, and they challenged the threshing machine company's right to use the name "Case" on plows and other tillage equipment. In 1916, the Wisconsin Supreme Court ruled in favor of the J. I.

Case Plow Works, forcing the threshing machine company to sell its tillage equipment without the Case name. This did little to boost the plow works, and after about 12 more years of slipping market share and financial difficulties, the J. I. Case Plow Works was sold to Massey-Harris. This left the threshing machine company free to use the name Case and to continue to develop its line of tractors and implements.

Antique Tractor Tip

Archival Knowledge
Tractor company archives were saved and are now housed with third parties. For example, much of Massey-Harris archives were given to the Ontario Agricultural Museum. For a small fee, these museums will perform research, and provide reproduction of company documentation.

Models to Consider Purchasing

Many models of Case Tractors make excellent work tractors. Models such as the lightweight V Series, or the heavier-duty D and S Series, and ultimately the Hundred Series that replaced them, all make great work tractors, especially if the tractor features Case's copy of Ferguson's system, the Eagle Hitch. While the Case Eagle Hitch in the early years didn't have true draft control, it was nonetheless a handy three-point hitching system compatible with modern three-point implements. The Case V and VA tractors in particular are excellent work tractors that are available in any configuration (row crop, utility, and so on). The Hundred Series Case diesel engines, such as the 500 and 800 Series, are excellent workhorses that have survived the years well.

It is not difficult to find a Case tractor that will work the fields while remaining a collectible over the years. The Case C, CC, L, and R Series and all their variations are great drawbar and belt pulley

tractors that are considered collectible. The Model C in particular has won a reputation as one of the sturdiest tractors built and was one of the most thoroughly tested tractor designs of its time. While there were some problems with earlier years of the CC gooseneck front-axle extension, most of the CCs held up well in the field, also.

As for machines whose highest and best use is being part of a collection, the earlier cross-mount Case tractors win hands down. These were created in response to demand for a more nimble and maneuverable tractor. These were made until 1927, and they have a solid reputation as a collector's item whose value appreciates. Of course, the unusual variants of the more common younger models also can be excellent collectible tractors and you should not overlook the opportunity to buy, say, an orchard version of a Case VA. The older Case two-cylinder gas engine tractors also have a very strong collectible reputation, but

these tractors are in very short supply and are expensive. The 12/25 model was the last Case two-cylinder tractor made, and this and the 20/40 model are both popular and valuable as collector's items.

FORD

Henry Ford, the son of a farmer, had turned his vision of an affordable, well-made car into a multimillion-dollar company with the Model T. As a child, he knew the drudgery of farming with horses, and vowed to apply the same vision he had for automobiles to agriculture. His designers set out to design a tractor that would suit the legions of small farmers, yet remain affordable and reliable. His designers developed a strong low-cost unit that seemed to meet all the needs of the small farmer. The only thing missing was a name. Since there already happened to be a tractor company named Ford, Henry Ford had to be content with naming his new tractor the Fordson, and the

Many antique tractors were available as industrial models, or specialized industrial models were made around the same time period. This 840 John Deere was built in cooperation with Hancock and based on the 830 chassis and some 830I parts. It was usually used with an elevating scraper.

company was named Henry Ford and Son. The company wasn't chartered until 1917, but Henry had nonetheless been involved in experimental designs and testing of farm tractors since 1908. Pundits in the tractor industry scoffed at Ford's tractor, but it became an immediate success. Henry Ford and Son rapidly became the second-largest manufacturer of tractors (after IH) in the United States. The success was short-lived, however, as a combination of the economy, competition from IH, and the need for every penny of spare capital to tool up for the production of the Model A car brought about the dissolution of Henry Ford and Son and the end of Fordson production in North America. The Fordson, however, continued to be marketed and

made in the United Kingdom for the English market, and a small number were imported to the United States.

The tractor industry still appealed to Ford, however, and in 1938, after seeing Harry Ferguson's three-point hitch and draft control system, Ford entered into an agreement whereby he would make the tractors with Ferguson's three-point hitch and in return Ferguson would sell and service the tractors. The tractors would be marketed through the Ferguson-Sherman dealer network and later through Ferguson Inc. This agreement was unfortunately verbal, and a parting of ways occurred in 1946. Henry Ford II formed Dearborn Motors Inc. in 1947, and Harry Ferguson sued the company and its officers for $251 million. Extended litigation that

lasted until 1953 resulted in an out-of-court settlement for a little over $9 million. At that time the three-point hitch was available to all manufacturers without license or royalty. Ford continued to make tractors until the late 1990s when New Holland, which began acquiring significant interest in the company in 1990, assumed total control of the majority of Ford tractor operations. The tractors today are now known solely as New Holland.

Models to Consider Purchasing

Ford made a great many tractors, and just about any model makes a great addition to a working farm. The Ford 600 and 800 Series of the 1950s, and the 2000 from the 1960s, are the best work tractors, but other models from the same time period, such as the 941 and the 6000, should not be overlooked.

For a cross-purpose tractor, sheer numbers recommend the Ford 8N, but the best cross-purpose Ford was the Jubilee, or more officially, the Ford NAA. The NAA offers many improvements over the 8N, such as a live hydraulic pump, and it is also more collectible. The 8N remains a very capable work tractor that has a certain collectible value too.

The Ford 2N and 9N share much of the same capabilities and design characteristics as the 8N, but these models probably would best be considered collectible tractors. Especially the 2N, with its steel wheels and other World War

II shortage-induced design changes. Other collectible Ford tractors are any of the English-built Fordsons, the Fordson Major and Dexta Diesels, and the early American Fordson. Also, don't forget any unusual variation of the N Series, such as the aftermarket I6 and V8 engine retrofits. These are quite collectible.

ALLIS-CHALMERS

Allis-Chalmers, a huge industrial giant by the time the company sold out in 1979, began when Edward Allis bought a manufacturing facility at a sheriff's bankruptcy sale for a mere $25 and some change. In 1901, through the merger and acquisition of several companies, the firm became formally known as the Allis-Chalmers Company. Allis-Chalmers began making its own tractor in 1915. This was the model 10-18, though Allis-Chalmers was at the same time pursuing a joint venture with a Swiss firm that made a rotary cultivator, or tiller. The machine was a tractor and tiller together and apparently not many were made or sold. Allis-Chalmers also attempted to market a "Track-Tractor" that saw very limited success. This vehicle was a "half-track," as the vehicle had tracks like a crawler but front wheels like a car or truck. The 10-18 became Allis-Chalmers' best-selling product, though total sales were low. Additional models were brought to market and additional improvements were constantly being added.

Many firsts in the tractor industry took place at Allis-Chalmers. In addition to the

In the late 1950s and early 1960s, several larger tractors were equipped with this large GM diesel engine. It has a reputation for being loud and oily but very strong. Massey-Ferguson model 98s and others with this engine often dominate the stock antique tractor pulls in their weight classes.

experiments with half-tracks, rotary cultivators, and advanced engine design, the first tractor available with rubber tires was the Allis model U in 1932. Allis-Chalmers also experimented with duplex tractors, set land speed records with its rubber-tired tractors, and later on led the industry in experiments with fuel cell technology. In addition, common wisdom has it that Allis-Chalmers tractors exhibit, on average, the best power-to-weight ratio of any line of tractors from any manufacturer. Of particular note in the history of Allis-Chalmers is the company's purchase in 1931 of Rumely, the esteemed tractor manufacturer responsible for the successful Oil-Pull tractors of the 1910s. As late as 1935, equipment with the Rumely name was still available

through Allis-Chalmers, though it is believed that the company produced no Rumely equipment after the purchase.

Models to Consider Purchasing

Many excellent antique tractors were made by Allis-Chalmers. Any of the D Series and the later 100 Series tractors (e.g., 170 and 190) make excellent work tractors that are complete with, or easily converted to, a three-point hitch. Be aware that the move to the three-point hitch was sporadic through the D Series line, and many tractors, especially the earlier ones, either had a live but nondraft-sensing hitch, or they had the A-C snap-coupler hitch, which is a lower link sensing draft system that was incompatible with three-point implements. Various changes were made to the D Series tractors and these changes were known as Series I, Series II, and so on. The Series II and latter tractors all come with independent PTO and all—except the D-17, which didn't get it until the Series IV changes—were available

Tractor tires have evolved into various tread designs over the years. Most of us are familiar with the cleat designs in wide use today, but the continuous tread design seen here was common when these tractors were new. Restorers looking for original style treads will be disappointed to learn it is hard to find a set like this. A few of the older designs are being reproduced, and some of the suppliers in the appendices can help you find these designs.

with a three-point system. The larger D-21, which broke the 100 horsepower barrier for A-C, and the 170, 180, and 190 serve best as large ground tillage machines. Be aware that many A-C tractors had excellent power-to-weight ratios and are popular with tractor pullers and others for competitive uses for antique tractors. Several A-C models make great cross-purpose machines. The WC, WD, and WD-45 are excellent narrow front midsized tractors that do well with collectors and can pull their weight on the farm. The latter model of the WD-45 also had the snap-coupler hitching system. If snap-coupler implements can be found, this model is very handy in the fields. The smaller brothers to the W Series, the B, C, and CA, are excellent collector's machines and can fill niche roles around home, farm, business, or estate. The model B is particularly popular as a mowing tractor when retrofitted with a modern mowing deck, such as those available from Woods. In fact, the model B was highly regarded by other tractor manufacturers and provided stiff competition for International Harvester and John Deere.

As for collectibility, any of the older Letter Series, especially the rarer U and A machines, would make excellent choices. The U was the first tractor in history to sport pneumatic tires. The G is a very unusual small tractor that's well suited to cultivation and holds its value very well. Like any make of antique tractor, any unusual variation of more common models, such as industrial variant,

"potato" specials, and others are all excellent choices for collections. The older models denoted by horsepower ratings (e.g., model 18/30) are highly collectible and should be considered.

MINNEAPOLIS-MOLINE

In 1929, just a few months before the stock market crash, Minneapolis-Moline was formed through the merger of the Moline Plow Company, the Minneapolis Threshing Machine Company, and the Minneapolis Steel and Machinery Company. Moline Plow provided the tillage equipment and the Universal Tractor, and Minneapolis Threshing Machine provided the Minneapolis line of tractors, including the 17-30 Type B. Minneapolis Steel and Machinery brought to the merger the highly regarded Twin City Tractor line, which continued to be sold under the Twin City name until at least 1938.

Minneapolis-Moline grew and prospered during the next three decades. While never a very large tractor manufacturer compared to John Deere, IH, and others, Minneapolis-Moline nonetheless enjoyed a good reputation and put forth many excellent tractors, some quite innovative. It also was the manufacturer of one of the most collectible antique tractors: the MM UDLX, a tractor with a cab that served as a cross between a farm tractor and motorcar. In 1963 the White Equipment Company, which had already acquired Oliver and Cockshutt, acquired Minneapolis-Moline as a wholly owned subsidiary. In the early

1970s, Minneapolis-Moline and the others ceased to exist as individual brand names when White began manufacturing tractors all under the White nameplate.

Models to Consider Purchasing

Minneapolis-Moline made several excellent work tractors, the 335, 445, the 4 Star, and the 5 Star, and their variants all provide excellent characteristics for field work. Live PTO, live hydraulics, and a three-point lift make them well suited for today's implements and conditions. One of the larger tractors in Minneapolis-Moline's late 1950s lineup, the Moline Gvi and G1000 were large workhorses capable of providing tremendous horsepower to both the drawbar and PTO. Interestingly, the G 705, a later G Series tractor, found its way to the Massey-Ferguson company and also appeared on the market as a Massey-Ferguson model 97.

Minneapolis-Moline made some excellent cross-purpose tractors. Primarily the U and the Z models of tractors and their variants make good, strong, solid collectible tractors that can also be called upon to help out with the chores. The B. F. Avery Model R (originally introduced by Avery as the Model A) that Minneapolis-Moline acquired when it bought out B. F. Avery gave Minneapolis a very nice small tractor that would suit any owner as a cross-purpose machine. The company called this model the Minneapolis-Moline model BF. In fact, some of the pictures in the rebuilding

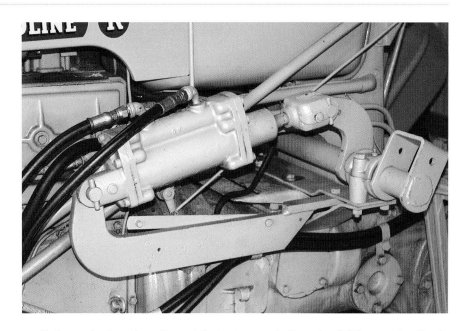

Until the agricultural equipment industry settled on the fairly standardized modern designs for attaching implements and hydraulic systems, all tractor manufacturers experimented with their own ideas. Here is an example of a hydraulic ram on a Minneapolis-Moline R that is driven from a live-powered pump driven by the engine. Many manufacturers, including Massey-Harris and Allis-Chalmers, developed some sort of variation on this design.

section of this book come from the restoration of a Minneapolis-Moline model BF.

As for purely collectible tractors, a Twin City tractor would be an excellent addition to any tractor collection. The Minneapolis model 17-30 types A and B are a good choice for any collector. From the Moline Plow Company, the Universal Tractor is a collectible. Of course, one of the most collectible tractors of all time, the Minneapolis-Moline UDLX Comfortractor, is rare as hen's teeth and highly collectible.

CATERPILLAR

The Caterpillar Tractor Company was the result of a merger in 1925 between the Holt and Best companies, both of California. Both had experimented, to varying degrees of success, with

tracklaying technologies. While Holt is credited with being the first successful manufacturer of crawlers, Best was also successful in this regard. Interestingly, both were developing the technology for different circumstances. Holt was refining tracklaying for soft peat areas of agricultural land, and Best was developing the technology for woodland logging operations where steep inclines and very soft soil call for the high flotation of tracks. The two companies had crossed paths once before in 1908, when Best had sold out to Holt. Daniel Best's son organized a new crawler company and their family was back in the business and began competing with Holt once more. At the second and final merger, the Caterpillar Tractor Company was formed.

One of the most significant advances in Caterpillar history

Antique Tractor Tip

Plow the 'Net
The Internet offers plenty for
farm tractor enthusiasts,
from photo archives and
discussion groups to parts
sources and rebuilding services.
Use it as a research tool.

was its introduction of a diesel-powered tracklayer in 1931. Caterpillar made diesel power a priority after a diesel-powered cable plowing rig in Africa outperformed a Caterpillar rig. Art Rosen, Cat's main diesel engineer, spent several years and over $1 million on the project. When it had a diesel model ready, Caterpillar deployed it to select customers for a trial run. No problems surfaced, so the company began general sales nationwide. Differences in fuel qualities around the country, however, led to problems, and Caterpillar had to search frantically for a solution. With some help from Standard Oil, a new diesel fuel was invented and Caterpillar's diesels led the way for the diesel revolution in the heavy equipment industry.

Models to Consider Purchasing

Any of the latter D Series Caterpillar tracklayers make great work units. Do be careful when purchasing Caterpillar tractors, as many still have the older style cable-lift blade systems if they have a blade at all. These cable-lift systems can be quite dangerous. If the tractor does not have a blade, be aware that hydraulically powered blade lifts similar to modern crawlers were not available at the factory for most of the older D Series and must be shop made and installed. The agriculturist will be annoyed by the lack of three-point hitch systems. Caterpillars are drawbar-only tractors. Good collector models include any of the Best and Holt units, in addition to the Caterpillar models that were Best Units carried over into Caterpillar production (models Thirty and Sixty). Any of the pre-D Series numbered tractors would make fine collectible tractors, but in particular you would want to look for a model Ten, Twenty-Five, or Twenty-Eight or any of the early diesels.

OLIVER

The Oliver Corporation formed in 1929 from the merger of four independent companies: the Oliver Chilled Plow Company, Nichols and Shepard Company, Hart-Parr Company, and the American Seeding Machine Company. All were long-established agricultural manufacturing companies with strong product lines. Oliver Corporation was immediately a strong player in the tractor manufacturing industry. The expertise and resources needed to develop and maintain a strong line of tractors came from the Charles City, Iowa-based company of Hart-Parr, and the 80-year-old Nichols and Shepard Company. Both companies had a very strong tractor manufacturing history. The Oliver Chilled Steel Plow company and the American Seed Company brought a full line of implements and equipment that could be sold with the tractors. Another milestone in Oliver's development was its acquisition of the Cleveland Tractor Company. This acquisition gave Oliver a model for the crawler market, and the crawler operation continued until several years after White bought out Oliver.

Hart-Parr had the distinction of building the very first commercially successful gasoline-powered traction engine. Its No. 1 model, built in 1901, was rated at 17 horsepower at the drawbar and 30 horsepower at the belt pulley. Much evolution and design correction took place over the next several years as the Hart-Parr product lines expanded. The other large tractor company in the Oliver merger, Nichols and Shepard, also had a very strong line of both steam- and gasoline-powered traction engines. While the company's foray into gasoline power was later than Hart-Parr's, its tractors were well designed and its previous steam models were very highly regarded. The combined history and resources of Oliver's four component companies made it one of the most talented and storied agricultural companies of its time. While just a scant 30 years later in chronological time, it was a technological lifetime later that the Oliver Corporation was bought in 1960 by the White Motor Corporation. White continued selling tractors under the Oliver name until the early 1970s.

Models to Consider Purchasing

Oliver made many great tractors well suited for work purposes, and the best of the bunch are the Hundred Series, such as the 880, 770, 550, and so on. These tractors have all the modern conveniences, such as live power takeoff and hydraulics, and three-point hitching systems were optional. Diesel engines became more common with the Hundred Series. The Thousand Series tractors, made through the 1960s, such as the 1600 and 1800, are also excellent work tractors.

Oliver also made great cross-purpose tractors, such as the early models (model 60, 70, 80, and 90). The newer Fleetline tractors, which include the models 55, 66, 77, 88, and 99, are great cross-purpose tractors, and many were available with the new Hydra-lux hydraulic lift when it became available in 1949. Other excellent cross-purpose tractors include Super Series (e.g., the Super 55, Super 66, and so on). These

While we all think we know what a generator looks like, this is an unusual one. This early tractor has a direct-drive, 6-volt generator for charging the battery. It is driven from the timing gear that also drives the distributor.

tractors, while at home in any collection, are field ready with live PTO.

Many tractors from Oliver's history are collectible. Any tractor from the premerger days, such as the Hart-Parrs, Nichols Shepards, and Cleveland Tractor Company's Cletrac crawlers are definitely collector's items. Tractors from Oliver's early history, such as the 18/28, are collectible as well. One of the more collectible—yet also younger—tractors in Oliver's product line is the 440, an unusual offset cultivating tractor of which few were made. It would behoove the collector to look for a 440 if considering a tractor from the Oliver line.

Other Makes

There are literally dozens of makes of tractors that I consider worthy of inclusion in this chapter. Other fine makes include Cockshutt (also marketed under the names Gambles Farmcrest and Co-op), David Brown, Long, Porsche (yes, the same company that makes sports cars), and Fate-Root-Heath. All have unique and interesting histories that I encourage you to read up on and research. Because most of these companies were small and the production numbers were limited, many of these unusual brands are among the most collectible. A full discussion of these brands is beyond the scope of this book, but

While the line of tractors named Farmall is a brand even those only vaguely familiar with tractors would still have heard of, few folks outside of the antique tractor hobby realize that International Harvester also marketed another line of tractors under the International badge. Most models tended to be standard tread models like this one and were medium to large tractors.

Antique tractors are great at helping out on the farm by pulling grain wagons. Here a John Deere B is bringing in soybean from the field.

I encourage you to borrow or buy any one of the number of books available that fully describes each manufacturer and its models.

Some of the makes you should look into:

▶ Cockshutt: The model 30 from the late 1940s featured the first commercially successful use of the live PTO. These are excellent tractors.

▶ Fate Root Heath: The Silver King line has excellent collectible tractors.

▶ Porsche: This unique machine is unusual in the United States and is collectible. The smaller models have an air-cooled diesel engine.

▶ David Brown: The maker of some very fine tractors, its diesel models in particular are well liked and respected in the field.

▶ Graham Bradley: One of the more beautifully styled tractors, it was the first and only large tractor sold by Sears, Roebuck and Co.

▶ Long: A small independent manufacturer of tractors that began making tractors in the post-World War II boom years and still makes them. Long has always had a respectable product offering manufactured in my home state of North Carolina.

In addition, there are several European makes, such as Lanz and Field Marshall, that were never widely distributed to North America and therefore are not discussed here. This section should give you some feel for the major brands and for the history of the antique tractor industry. There are many books on the history of the tractor industry in general and on the history and product lines of various manufacturers. I encourage you to contact Motorbooks International, the publisher of this book, for a complete list of all the titles it has available. 🚜

CHAPTER 3
Buying an Antique Tractor

To some, shopping for an antique tractor is the most exciting part of the experience of owning it. Unfortunately, many buyers purchase antique tractors before they learn enough about them, and only later does the new owner discover that the tractor doesn't meet his or her needs. While these are not necessarily impulse purchases, they are usually uninformed buys. This chapter will set forth the important considerations surrounding the purchase of an antique tractor. I encourage you to do additional research as well to be sure you know what you want from your tractor and that the model you choose will meet your expectations.

Typically, the purchasing process unfolds something like this: The buyer, while harboring a notion that there is a need for a tractor around his or her farm, home, or business, happens upon a tractor for sale. That tractor seems to address the needs the buyer can think up at the time, though the laundry list of needs hasn't been finalized. Attempts to learn more about the tractor are often frustrated by the fact that additional information about the model is

Straight, clean design; half moon hitch; and the articulated steering arm are the hallmarks of the Allis-Chalmers model C narrow-front tractor. Smaller than the W and D series that followed in later years, its size and capabilities suit it to smaller chores around the farm rather than heavy field work. Today, it is a popular model for hobbyists and restorers.

High crop tractors are typically very collectible and command premium prices. The prices these tractors fetch dictate that this model A high crop is destined for a life in a collector's barn and parades.

This Case model S that was restored by Gene Dotson is actually an orchard conversion and is a collectible example of a Case tractor. When belted up to a thresher at a farm show, it runs out great and has plenty of reserve power.

difficult to find. The asking price seems fair, so the buyer makes the purchase without any real understanding of the tractor or his or her needs.

In this scenario the buyer usually ends up with a tractor that is not ideal and may even be unsuitable or dangerous. The problem is that the buyer didn't fully understand his or her own needs for buying a tractor, the market for antique tractors, or the specific tractor he or she was buying. If one of your tractor purchases fits this description, you are in good company. Many owners of antique tractors, including me, have made the same mistake; sometimes it even occurs after you have owned a few tractors. That is because needs and motivations are vague and changeable, and there is often much misunderstanding and misinformation about antique tractors. It is much too easy to buy a tractor that is not close to your ideal tractor.

Information is the key to making a sound purchase of an antique tractor. If you find yourself reading this section before buying a new tractor, you are in an excellent position and should have a strong likelihood of buying the tractor that suits you and your operation very well. If you are not so lucky and have already purchased a tractor that you are not happy with, consider selling the tractor (most antique tractors depreciate very little and most appreciate), then read this chapter, and try your hand again at buying the tractor of your dreams!

Now is the time to start outlining what you need in a tractor. Plan on sleeping on the list for a few days before ever looking at a tractor. You should use this opportunity to read books, visit friends and co-workers with tractors, and visit antique tractor shows that are held routinely in hundreds of towns across America each year. This is how you will learn about all the different makes and models available. This process will also put you in contact with additional sources of information that can help you better define just what the dream tractor will be like and what types of equipment options and implements you may be interested in. From here I will outline the basic issues of selection. Once through that, you should be able to draw your own conclusions as to what type of tractor should be best for you.

Ford high-crop tractors fall into the collectible category, and most folks buy them to show rather than to use. This one is in excellent original shape.

Acquiring Your Tractor with a Plan

After information, the next two most important tools you have to help select a tractor is a plan and patience. The plan should outline all the things you know you must consider in a tractor and how you will evaluate those things in each tractor you see. This plan should detail things like what types of manufacturers you will consider and what types of options and systems should be present on the tractor. This plan shouldn't be written in stone, but should be a guideline that evolves as you look at prospective tractors and make further decisions. The understanding needed to develop this plan can only come from studying this book and others like it, as well as visiting those more experienced with antique tractors, attending antique tractor shows, and the buying process itself.

You will learn more about tractors with each one you visit. For this reason, you should be patient while shopping for one. You will need the patience to wait for the right model and make of tractor, the patience to wait for the right instance of this model and make, and the patience to look at as many tractors as possible (not to mention the patience to wait for the right price!). There are, of course, other things you need to keep in mind: Take notes when you audition a tractor. If possible, give the tractor a thorough workout. This section will take you through the decision-making process, helping you to decide which tractor is right for you.

Your Needs

Before you pick up the newspaper and start calling about tractors, take the time to sit down and ask yourself a few questions. What types of work will I be doing with this tractor? What implements will I need to buy and how expensive and available are they? What is my price range for the

When buying an antique tractor, it's advantageous to have an original, legible serial number plate present on the tractor. Some models of antique tractors were prone to having the plate knocked loose or losing its legibility from damage, corrosion, or fading. This makes model year identification difficult unless you have studied that particular model of tractor. Having the original plate still affixed is a requirement if you are looking to purchase a highly collectible and valuable tractor.

tractor and required implements? What size and power class of tractor should I be considering?

Be careful to anticipate your current needs as well as your future needs, as these will heavily influence what type and size of tractor you should buy. If you plan on using this tractor for work, the implements you plan to use will drive your decision. If your motivations and needs tend toward collection and restoration, then historical considerations, collectibility, and nostalgia may more heavily influence your decision. Simply outlining these needs and motivations will go a long way toward helping you decide on the right tractor for you.

If you are planning to use your tractor for large jobs around your property, these needs will override your other uses, though these other uses may be more typical. In this case, you should buy a tractor sized to handle the larger implements. For example, let's assume you are using the tractor mainly for light property maintenance, and you also plan to raise a few acres of hay. Almost any small antique tractor will handle all the property maintenance chores. These small tractors will also run mowers, conditioners, rakes, and tedders necessary for putting up hay. You will need a midsized or larger tractor,

however, for pulling a PTO-powered hay baler. Be sure to consider all types of work when defining your dream tractor.

Configuration: Different Styles of Tractors

ROW CROPS

A row crop is any tractor whose front and rear wheels and axles are designed to maximize the farmer's ability to work in the fields among crops that are grown in rows. People often mistakenly equate row crop tractors with tricycle tractors—those with one or two small wheels set under the front of the tractor. Although tricycle models are row crop tractors, a tractor with a traditional front axle and widely spaced front wheels can also be a row crop. The defining characteristics are highly adjustable axles and wheels and a stance that is sufficiently high to allow the tractor to pass over crops.

Tricycle tractors have a few quirks to be aware of. Primarily, you want to make sure that the front wheels, if they were meant to ride up and down independently, do so. Be aware of any pedestals that have been broken and welded. The front end of any tractor takes a lot of abuse, and the narrow front tractor is especially prone to weaken under hard use. On steel wheel tractors, be sure that any rust on the front wheels is only surface rust and not something that threatens the strength of the wheel. When operating a tricycle tractor, be aware that its hillside stability is typically less (though not as bad as one would assume), and if the distance between the rear axles and the front wheel(s) is short, the danger of a rearward flip is increased.

STANDARD TREAD, WHEATLAND, AND WIDE-FRONT ROW-CROP TRACTORS

All of the above tractors share one thing in common: They have four wheels and the front wheels are spaced widely apart at the ends of a front axle. These tractors also have some significant differences. Utility tractors often do not have adjustable wheel spacing, while the wide front-end row-crop tractors will. Wheatland tractors will not have adjustable wheel spacing, but are without exception very large tractors, and more likely to have diesel engines and less likely to have auxiliary systems such as lifts, remote hydraulics, and so on. The wide-front row-crop tractors will have the highest ground clearance and hence the highest center of gravity and less stability. A typical utility tractor is the Ford 8N, a typical wheatland tractor is the John Deere 830, and a typical wide-front row-crop tractor is the Farmall A.

These tractors can have many different configurations. The Farmall A mentioned above is an offset tractor, meaning its engine and drivetrain are offset to one side. That way the operator does not straddle the transmission when operating the tractor. The transmission and the engine are to the left of the operator and nothing but the ground is below

When buying a tractor, try to mitigate your expectations to its cosmetic condition. Paint is easy and fairly cheap to apply. Intractable mechanical issues and incorrect or missing parts can be expensive. While this tractor doesn't shine with a new paint job, it is actually a nice tractor. It runs well, the sheet metal and the radiator are in good shape, the steel wheels are solid and correct, and the bird's nest on the exhaust is probably free with the purchase of the tractor!

the footrest. This gives him an unparalleled view of the row crops he may be working with. This design, while credited to International Harvester for its use in the Farmall A, has also been used by other companies, such as Oliver on its 440 model and Ford on its 541 (only operator seating is offset in this design).

Another configuration worth mentioning is the industrial configuration tractor. This design is usually a variant of a company's utility model of tractor. Further modifications are made to accommodate industrial implements and production systems. Other modifications typically include larger front axles, different wheel sizes and tire options, operator platforms and layout, color scheme,

and the lack of typical agricultural options such as a belt pulley. These tractors are usually used as tugs, fork lifts, moving power units, and chore vehicles.

CRAWLERS: DON'T OVERLOOK BUYING THIS CONFIGURATION

What do I need a bulldozer for? That's the question I hear every reader asking right now and it's a legitimate one. A bulldozer and a crawler, however, are not necessarily the same machine. A bulldozer is specifically any tracked vehicle that has a large front blade for pushing and leveling. Most antique tracked vehicles were never fitted with a 'dozer blade, nor was one typically available from its maker. These antique tracked

Allis-Chalmers came out with the model G as a small garden and truck farm cultivating tractor. Excellent visibility and narrow wheels and tires made it an excellent tractor for this work. The G is one of the most popular models among Allis-Chalmers collectors.

vehicles, or crawlers, often were true agricultural machines, having a PTO shaft or a belt pulley, and in rare instances, a three-point hitch.

Antique crawlers were designed strictly as agricultural machines and have some significant advantages over wheeled tractors. Typically they do not compact the soil as much as wheeled tractors, they are much less likely to get stuck in fields, are usually more stable on hillsides, and will outpull a typical wheeled tractor in the same power and size class. They have a few detractions too. They require more maintenance, and repair of the drivetrain, or undercarriage, can be costly. They are much heavier than the same class of wheeled tractor and cannot be used on concrete or asphalt.

Age

How old should your antique tractor be? Generally, you should consider any age, but the post-World War II tractors are more plentiful, and if you plan on using the tractor for extensive work, these will be the only ones to have the systems and options needed for modern production. If you are purchasing the tractor for a collection, then any age should be considered and many collectors feel that "older is better." This is a matter of personal preference, of course. The very old tractors usually appreciate faster, are often a load of fun to own and restore, and can be used with any drawbar implement in modern production settings. Just keep in mind that the older tractors will be more difficult to find parts for, to repair, and to adapt to modern farming.

Size

"What size tractor should I buy?" will be your next question. Size, as it relates to physical measurements, should only be a minor concern. More important than dimensions is the tractor's horsepower rating. If you don't have the power you need, the fact that your tractor is a convenient size will not be much of a benefit. Antique tractors were designed to provide a certain horsepower, and the actual weight and physical measurements were by-products of the design process. For that reason,

Antique Tractor Tip

Backyard Jewels
Finding antique tractors to buy often means a slow drive down country roads, looking for old tractors abandoned in pastures and poking out of barns. If you spot an interesting possibility, approach the owner, express your interest, and leave a calling card so you can be reached if he or she decides to sell.

tractors of comparable horsepower may vary significantly in size. Determine your power needs first and then explore questions of size and weight. The published results of the Nebraska Tractor Tests will be helpful here. Here are some guidelines for determining minimum horsepower requirements of various implements:

▶ Small square-bale hay balers require 25 to 30 PTO horsepower; add 5 to 10 horsepower for moderately hilly work or heavy hay.

▶ Double-acting disc harrows start at 4 feet in width and require about 12 drawbar horsepower. Add 7 horsepower for each foot of width after 4 feet.

▶ Plows require approximately 10 horsepower per 14-inch bottom—more in hilly, thick soils; less in flat, sandy soils.

▶ Planters and cultivators have minimal horsepower requirements and any antique tractor can cultivate at least one row—more as horsepower increases.

▶ Moderately sized (around 8- to 10-foot) planting drills require at least 12 to 15 horsepower at the drawbar.

▶ Snow-blowers come widely sized with many horsepower requirements, but tractors giving 15 horsepower at the PTO are the smallest you would want to use for this application.

These John Deere 520s are both a collectible tractor and excellent chore tractor. The 520 was the successor of the model 50, which was the successor of the model B.

▶ Any tractor with live hydraulic pumps can power a front loader, and the loader will have to be sized by other concerns, such as front-end strength of the tractor, pump flow in gallons per minute, type and size of bucket, and so on.

▶ Tractors used to mow lawns should weigh less than 3,000 pounds, as damage to septic fields and drainage tiles may result.

When you are unsure what size of implement the subject tractor can operate, err toward

Antique Tractor Tip

When In Rome . . .
When buying an antique tractor for use, seriously consider buying a make of antique tractor that was most popular in your area during the time the model was new. This assures reasonable prices and ample local knowledge, as well as improved parts and implement availability.

Smaller John Deere tractors were also two-cylinder, but their arrangement was vertical instead of horizontal. This model 430 shows the same basic engine that was in use with its predecessor, the model M.

the smaller size. Some implements may be too large for your tractor, even if you have the horsepower to operate them. For example, huge 7-foot scrape blades can be pulled by many smaller antique tractors, but would be dangerous for the tractor to use because the large blade would make the front end of the tractor tend to tip up when lifted. Other issues, such as ground speed, will rule out other implements. Implements like tillers require very slow ground speeds. Most antique tractors are not capable of "creeping" like the modern tractors can and do not run tillers well. While there were aftermarket hi-lo transmissions for antique tractors, such as the Sherman transmission for the Ford N Series tractor, which make the tractor more suited to roto-tilling, finding a tractor

with this option may be tough or time consuming. Your local farm tractor implement yard will be able to help you determine the horsepower requirements of some of the implements you may be considering that I do not mention.

Narrowing Your Choices Down to Specifics

DECIDING ON BRANDS AND MODELS
Typically, choosing a manufacturer first is easiest. For example, some folks just cannot live without the sounds of an old two-cylinder John Deere, while others simply have to have the Ferguson draft system that existed on the Ford N Series tractors and later on the Ferguson TO/TE Series tractors. Others love the split-block engines

of Minneapolis-Moline tractors. Still others are partial to the simplicity of the Farmalls. And others love the power-to-weight ratios of the Allis-Chalmers, and the list goes on. Most of the large, reputable tractor manufacturers made good machines that were solid and reliable. Choosing the manufacturer usually comes down to less objective criteria.

One reasonable approach for narrowing your choices is to consider buying a tractor from the manufacturer that was most successful in your local area during the time period that interests you. This approach should provide you with the largest supply of used tractors to look at, as those are the tractors most likely to be found in the area. In addition, it also increases your likelihood of having neighbors and friends who know and understand your machine and can help you with it. It also should ensure a larger pool of parts and specialized implements. The tractors that were the most popular should be suited to the land and the agricultural uses in your area.

DECIDING ON THE MODEL WITHIN THE BRAND
Deciding on a model is more difficult than choosing a manu- facturer. At this stage you bring out the needs checklist you created earlier. This list of needs and the horsepower range they demand, and the configuration choice you have made, will drive the decisions for a particular model. For example, if I were purchasing a tractor and I intended to use it as a restoration

project that will be included in my collection, as well as for various light chores around my property, I would look into the small tractors made by International Harvester. IH was the most successful manufacturer in my area. This type of tractor would easily handle my needs for light chore work, take up little storage space in my collection, and be easy to handle as I repaired and restored it. It would also be easy to transport to antique tractor shows.

Create a list of one to three tractor models made by your chosen manufacturer. These models should fit your chosen age range, chosen horsepower range, and configuration. This is the core group of models you will shop for.

The smaller John Deere tractors, such as this model 430 and its predecessors models 40 and 420, are highly collectible and sought after. This is especially true if they have an unusual configuration, such as the high-crop variation. The model line that consists of the 30, 320, and 330 is also a line of vertical two-cylinder engine tractors of John Deere that finds its way into many collectors' garages.

Finding and Inspecting Tractors

SOURCES OF ANTIQUE TRACTORS

There are several ways to go about looking for antique tractors to buy, and your best bet is to try to cover all of them. I routinely check the newspapers, local weekly publications advertising tractors for sale, state agricultural newspapers, and more recently, the Internet. Finding tractors for sale at antique tractor shows is helpful too, as many of them will have been restored into top-notch shape. One nice aspect of shopping at shows is that you can often see the tractor under load at a show, which might feature demonstrations, tractor pulls, or a dynamometer. I also make sure that all my friends and neighbors know that I am on the lookout for a tractor. Many good tips come from them.

VISITING AND EVALUATING THE TRACTOR

Everyone should remember that there are three rules of thumb when buying tractors. These are in addition to your typical common sense rules regarding buying expensive pieces of equipment:

1. Never assume the owner is correctly identifying the tractor or its capabilities. The majority of the tractors I look at have been misidentified or misrepresented (usually unknowingly) in some way.

2. It will always take more time and money than you think to fix a tractor's problems.

3. If you are collecting, a complete tractor with good sheet metal beats a running but incomplete tractor with poor sheet metal. The opposite is true if you are buying the tractor for work.

The first rule is self-explanatory. Never in any market have I seen so many expensive objects being bought and sold with such little information. Very little of this has to do with malice or the purposeful withholding of information, it is just that so little information is known by the general public. Allow me to quote a few real-life examples. Within the last few years I have seen the following misinformation perpetuated by sellers in the classified ads or in person:

Junkyard tractors like this one, sporting bullet holes and looking quite rough, can be a parts bonanza. When looking for repair or replacement parts, be sure to thoroughly look over the tractor and pick off any serviceable parts you can afford to buy from the junk dealer. If you don't need all of the parts, they can be used as trading stock for parts and supplies you do need.

- ▶ Harrison Family Tractor. This was really a Massey-Harris-Ferguson model 22.

- ▶ Ford Red Belly. This was actually a Ford 3000.

- ▶ Ford 8N. It was really a 9N or 2N (this is a very common error that I have seen many times).

- ▶ Massey-Harris Pony. This was really a Massey-Harris-Ferguson Colt.

- ▶ "Rare Antique Tractor for sale." This turned out to be a homemade tractor that was indeed old.

- ▶ Many instances where the year of manufacture was not known or was wildly misidentified.

Again, most of these circumstances involved sellers whom I regarded as at least reasonably reputable and honest. The problem is that casual users and owners of antique tractors don't have and can't find any rock-solid information about their tractors. Sellers usually just end up passing along in classified ads the same misinformation they were given when they bought it or the best guess that they could come up with.

The second rule is also self-explanatory. Often a problem exhibited by the tractor is the result of a more fundamental problem that isn't obvious during brief inspection visits. My favorite real-life example is the brakes on an N Series tractor. Let's say the right brake doesn't work. You replace it. You notice a lot of oil and grease but dismiss it as the accumulation from 40 years of work. The new brake shoe holds for a grand total of three weeks. Upon disassembly, you realize the oil and grease is leaking fast; enough to ruin any lining you install. You replace the rear axle seal and the brake shoe again. This lasts three months. Upon disassembly, you realize that the axle is leaking again. You pull the seal and discover the old seal had worn a groove in the axle shaft and the new seal couldn't keep the oil from leaking because of the groove. You then purchase a sleeve for the axle, a new seal again, and a brake shoe for a third time. This repair runs hundreds of dollars when you initially calculated maybe $100.

Other times parts are so hard to find or are in such high demand that the repair becomes more expensive than you expected. For example, prices for tractor engine manifolds vary widely. It would be really easy to underestimate the price of a manifold if you do not research this explicitly.

The third rule is a mix of common sense and arcane knowledge. A "complete" tractor is

Antique Tractor Tip

Tanks Ain't Trivial
Carefully inspect the radiator when buying a tractor. New cores and tanks are very expensive, and replacement or accurate reproduction tanks may be nearly impossible to find.

important, even if you will not be restoring it. Locating and paying for original parts, especially "new old stock" (meaning the parts that were made by the O.E.M. (Original Equipment Manufacturers) for that particular tractor during the time the tractor was not obsolete), is difficult and time consuming. Finding a complete tractor with repairable parts is preferable. Likewise, with the sheet metal: It is very expensive to repair or replace, if you are lucky enough to find replacements. The collector also has time to fix and repair a nonrunning tractor, and it is typically not terribly expensive to do so, especially if you do it yourself.

The standard is different for the person buying the tractor for work. The work tractor purchaser just wants a tractor that is complete enough to service his or her needs, even if some nonessential parts are missing or replaced with homemade or off-brand items. The work tractor also doesn't need all its sheet metal nor does it have to be pretty. The tractor should run well, though, since a work tractor usually has chores ready for it when the purchaser gets it home.

When inspecting antique tractors, be sure of several things: Take notes, as little things are forgotten when you are making your decision several days or weeks later. Be sure to test drive the tractor and put the tractor under load to evaluate its performance. I like to hook the tractor up to a large wagon, plow, disc, or whatever I can, if possible. Almost any running tractor can move about easily and smoothly when there is no load. Under load is

Antique Tractor Tip

Three Rules of Tractor Buying
1. Never assume the owner is correctly identifying the tractor or its capabilities.
2. It will always take more time and money than you think to fix a tractor's problems.
3. If you are collecting, a complete tractor with good sheet metal beats a running but incomplete tractor with poor sheet metal. The opposite is true if you are buying the tractor for work.

when the tractor will disclose its pesky habits and problems. Loud or unusual noises need an explanation from the owner. I have found that most sellers will try their best to answer any question honestly, but will often not volunteer any information about the tractor without prompting, so ask a lot of questions. If a tractor has serious problems or inadequacies, you want to find that out before you spend your hard-earned money, not afterward. In short, go over it with a fine-tooth comb!

Major Concerns

TIRES

Rear tractor tires are incredibly expensive, as even the smallest tires start at $100 each plus mounting, tubes, tax, and so on. This actually should be one of the first things you look at. Do the tires have most of their tread? Are there any cuts in them? Tractor tires, because of the terrain and their proximity to heavy, sharp implements, are prone to cuts. The most common problem

with tires on an antique tractor is checking and cracking. These are caused by age and exposure to sun and can be dangerous or cause serious damage if they fail while the tractor is in operation. I would be concerned with a set of rear tires that has extensive checking and cracking, unless the tractor will be used for showing, collecting, and light use on level ground. Front tires, which cost in the $30 to $60 range, plus mounting and so on, are not as big an issue, but figure them into your costs if they are in poor shape.

SHEET METAL

Sheet metal can be a major factor in your decision to buy a particular tractor, or it can be just a small detail. If you are using your tractor for work, and the sheet metal at least performs its function (keeps branches and chaff out of engine and radiator area, fenders keep mud off you and keep you out of the wheels, and so on), then sheet metal may not be a concern. If you are using a tractor for collecting or restorations,

When looking at antique tractors, note any missing parts and try to get a sense as to how expensive it might be to replace them. A beginning buyer will be amazed to learn how expensive some parts are. An original NOS or used radiator cap, such as the one you see here for this John Deere, is quite expensive. To give you an idea, I can buy a piston for the engine cheaper than I can buy this radiator cap. By cataloging missing parts and adding these costs to the tractor, you can more easily compare true costs between the tractors you are considering buying.

however, the condition of the sheet metal will probably be the overriding concern. Sheet metal is very expensive to repair or replace, if replacements are even available. This is especially true of the "orchard" tractors and other similar tractors that had extensive sheet metal. If you are buying a tractor for show and collecting, you need to pay particular attention to the condition of the sheet metal. If you are using the tractor for work, just be aware of the importance of sheet metal to collectors. If you buy one with good sheet metal, you will be able to resell your tractor in the future for more money, and it should sell much more quickly.

Small dings, dents, and cuts in sheet metal are not a problem. Big dents are a problem, as are spots where rust has eaten completely through. Also be aware that sheet metal is often completely missing. Many tractors didn't have fenders, but many did and their absence can be costly. Small things like battery boxes are almost always rusted through, and others, such as toolboxes, are often missing. Check the condition also of attachment hardware, such as latches and braces, and look for rust underneath pieces of sheet metal. Reading the rust is helpful, as a solid coating of light rust on the sheet metal means the tractor spent several years of its life stored outside rather than in a shed or garage. Be aware of the small details and take notes. Between two otherwise comparable tractors, the one with better sheet metal is often the better value.

MECHANICAL
Engine

While some of this is obvious, other aspects of evaluating an engine are not. Of course you want to buy a tractor that starts right up and runs smoothly, but often these tractors have been sitting unused for quite some time and were maintained by people who were better farmers than they were mechanics. You have to anticipate that some of these tractors will not run like tops but will still have good, solid engines. It can be difficult to judge the engine by sound alone, as the engines in antique tractors often sound different than the engines most buyers are used to hearing on a day-to-day basis. If the engine seems loud and runs rough, you may just be listening to an engine that needs some new spark plugs or has a straight pipe for a muffler. Or it might not be running on all cylinders because of a burned exhaust valve. I bought a tractor once that would hardly start and ran rough. A quick inspection indicated the spark plug wires were routed to the wrong spark plugs. Then again, I ran across another tractor once

Antique Tractor Tip

Tires=$$$$
When considering a tractor for purchase, closely inspect the wheels and tires. Both are expensive to replace, and wheels can be expensive to repair.

with the same symptoms and the engine was just completely worn out and would not start well because the engine had almost no compression.

The results of three tests that report on an engine's vital signs are very handy to have. The first is a compression test. You want to see an engine whose compression readings are within 75 percent of factory specs and each reading is within 20 percent of the others. A variation on the compression test that you may want to make is a "leak down" test. This test will tell you not only what your compression is, but what is causing any loss of compression. You also want to test oil pressure when the engine is cold and when it is hot. A reading that deviates much from factory specification is not good. The last test is a vacuum test. This will evaluate the condition of the manifold and manifold gasket. Check the later part of this book for more information on these tests.

Here is a checklist of things that indicate a poor engine—or at the very least, indicate that the engine should be checked carefully

- Presence of oil in the coolant water

- Presence of coolant water in the engine oil

- Bluish-black smoke from the exhaust or the engine breather cap

- Water or oil leaks from around the head gasket

Antique Tractor Tip

Junkyards Are for Junk!
When selling a tractor, never sell to a scrap metal dealer regardless of condition. The vast majority of the time you will make much more money selling the tractor for parts. Advertising in the periodicals and Web sites listed in the appendix of this book will help you to find buyers for the parts.

- Cracks in the engine block or head

- Clicking or knocking noises from the engine

- Engine will not idle smoothly but passes vacuum test

- Excessive gas leaks at the carburetor

- Very noisy valvetrain

- Lots of black smoke in the exhaust. This is especially bad in diesels

- Excessive oil leaks from the rear of the engine

- Inability of the engine to maintain a constant engine speed

- Bubbling in the radiator when the engine is running

Drivetrain

The drivetrain is difficult to evaluate because it often requires

heavy loads to trigger the occurrence of the problem. Of course, the tractor should move about with the minimum of noise, and under no circumstances should a clattering or knocking be heard. Hard hammering by the transmission indicates broken gear teeth; loud, higher-pitched noises indicate bad bearings; and a rubbing or grinding sound can mean a number of things, such as bent shift forks or loose or misaligned components. Some problems, such as jumping out of gear, seem serious to those unfamiliar with transmissions but often these problems are simple and cheap to fix. Refer any problems or noises in the tractor to the owner and simply ask point blank if he or she has a diagnosis for the problem. If not, ask an experienced friend or mechanic to help you evaluate the transmission if you are in doubt.

Do realize, however, that the gearboxes of antique tractors are usually worn and involve technology that is not as quiet as modern transmissions.

The Freeman loader was one of the most popular aftermarket loaders available when these antique tractors were new. Many of them have survived and are serviceable today. Before you find, buy, and install one, you need to realize your tractor should have power steering. Steering is nearly impossible with a full bucket without power steering. The rear end becomes very light with a bucket attached, and some sort of counter weight added to the rear is advisable.

A continuous whir or whine can be expected, especially at full throttle while in road gears, though it shouldn't be overly loud. Shifting should be reasonably smooth, and the shift lever should not be loose or wobbly in its mounting. You want a transmission that emits the minimum of noise and operates with a tight, smooth feel.

The clutch is next. Is it easy to engage and disengage? Does it engage smoothly? Does it release quickly and smoothly? Is there any chattering or vibration? Try to put the tractor through a load test to check for slippage. Any heavy load will do here, such as plowing or standing on the brakes if the brakes are in good condition.

This is a component that should operate smoothly and quietly, or expect to repair the clutch soon. Be sure also to test any secondary clutches used for live power accessories, belt pulleys, and so on.

Another drivetrain element on some tractors, and one that deserves consideration, is a torque amplification unit. This unit, such as IH's Torque Amplifier or Minneapolis-Moline's Ampli-Torc, lowers the tractor's final drive ratio for extra lugging power. It is common for this component to be in disrepair, which typically means the tractor will operate, but not in one of its ranges. Be sure to check it, as it can be more costly to repair than any other component in the tractor.

John Deere's hand clutches also need close inspection. These clutches often are quite worn and need attention right away. While typically it is easy and inexpensive to bring the hand clutch back into working order, some of the repairs can be costly. Also, the clutch can be a good diagnostic tool for determining whether the crankshaft (camshaft on the H model) has too much play. Look at the action of the flywheel and the clutch itself during engagement for any signs of excessive crankshaft movement or wobble.

Brakes

Unless the tractor you are looking at has been recently restored or had a brake overhaul, this is one item you can bet isn't working or isn't working to anywhere near its original performance level. The brakes are usually either a band system or a disc system operating on drums that revolve on the shafts leaving the differential. While many different designs exist, they all behave the same way. A linkage either pulls directly on a band, or it pulls on a lever that tightens a band or applies pressure to a disc pad. The linkage is operated directly by pedals that the operator uses.

The main problem will likely be that one or both brakes do not work or work poorly. If the tractor was used for plowing, the left brake, which is used more frequently than the right during plowing, will show greater wear. Brake problems can also arise from grease and dirt build-up on the pads. This build-up usually comes from transmission leaks. Brakes will also wear excessively if adjustments that compensate for wear haven't been made. Another culprit for unusable brakes is that the band/pad lining material is worn past specifications. The trick to evaluating brakes is to try them out, both in straight-way braking and in brake-assisted turning.

Auxiliary Systems
Three-Point Hitches and Hydraulic Lifts

When checking out a three-point hitch and the associated hydraulic lift, you want to see three things: no excessive leaks, a lift and hitch that is complete, and a lift that raises implements quickly and smoothly and does not drop its load once it is raised. A properly operating three-point hitch will lift the front end of a tractor off the ground before it refuses to lift a load too large. One that is properly operating also should keep an implement in the air nearly indefinitely. Realistically however, you should expect an antique tractor to lift properly sized implements smoothly and without hesitation, and it should keep that load in the air for more than an hour or two.

An additional issue is whether the system uses draft control or has "down" pressure. Early Ford tractors, and most makes after the mid-1950s, had a system by which the hydraulics of the tractor lifted the hitch automatically based on field conditions sensed by the hitching system. The drawback is that

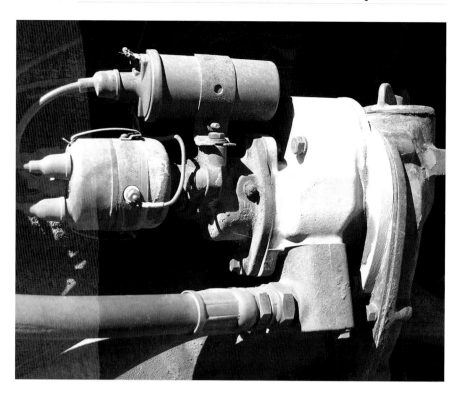

Farmall M tractors all had a belly hydraulic pump mounted in the frame between the clutch and transmission. Unfortunately they were not live pumps (they quit pumping when you pushed in the clutch) and were low-pressure, low-flow pumps. Farmall came along with an additional pump driven by the timing gears of the engine and mounted behind the distributor or magneto, as seen here. These were higher flow and higher pressure pumps better suited to loaders and many other hydraulic-controlled or -operated implements.

these early hydraulic draft-sensing lift systems do not have the capability to push down on the implement, they can only let the implement "float" down. Nondraft-sensing lift systems do have down pressure.

Remote Hydraulics

Many implements have hydraulic rams and motors that are driven from the tractor's hydraulic pump. These implements require that the tractor have "remote" hydraulics, that is, the ability to divert hydraulic pressure and flow to other implements and machinery. Common examples of this are front loaders, road tires, and axles for field tools that are raised and

lowered using hydraulic cylinders and implements with hydraulic motors such as log splitters. Some time spent now deciding how important this will be in your collection or farming operations will be beneficial. Many small farms get by with nonhydraulic implements and simply do not have a need for them. When they do run across an occasional use for one, some adaptations such as a PTO-driven hydraulic pump fill the void. Remote hydraulics are handy and became fairly common on tractors by the late 1950s.

The same inspection procedures used for hydraulics lifts and three-point hitches

Front-end loaders were popular attachments for antique tractors, and many still survive today and are being used. This International front loader is mounted onto a Farmall model M row crop and is probably the same age of the tractor.

Balers, mills, fans, shredders, and choppers, and just about all other forms of powered farm equipment, will accept power input from a belt or PTO shaft. Even if you are planning just to collect tractors, there are often demonstrations at shows using them, and you may want to be sure this feature works. When the belt pulley or PTO shaft is engaged, it should not wobble or emit strange or loud noises. Typically you should just hear a hum or a slight whir from the gearbox that drives the pulley. PTO shafts are typically very quiet and you should hear nothing unusual here. Neither should be leaking excessive oil. If you are looking at a tractor with a belt pulley and it is a "paper" pulley, be sure the paper is in good condition. Be sure the PTO shaft is the 1 3/8-inch standard shaft, or plan on purchasing an adapter. It should also have, or you should plan on purchasing, an overrunning clutch if the shaft is not live (see below).

apply to inspecting remote hydraulic systems. The system should be reasonably clean, with no excessive leaks, checked and cracked housings or hoses, and it should operate reasonably sized equipment and motors smoothly and without hesitation. Look for healthy and clean couplings and for signs of regular servicing.

Power Takeoff and Belt Pulley

The power takeoff shaft and the belt pulley are important if you plan to power auxiliary equipment.

Live versus Non-Live

Often auxiliary systems are live; that is, the PTO shaft and hydraulics were designed to continue operating, even if the operator disengages the clutch. To remove power from the auxiliary system, you press the clutch in further, or use a separate hand clutch. If the tractor has live power, make sure that the clutches correctly operate this feature.

Specialized Equipment

These are implements that are bought with the tractor and mounted to it in such a way that

Antique Tractor Tip

Let's Make a Deal
If a seller is particularly stubborn about a price, ask if he or she has any matching implements to throw into the deal. Sellers that don't budge on price may be willing to include implements or delivery (both valuable additions to the deal) without a change in price.

the tractor usually is dedicated to that use. Examples of such equipment include front loaders, field sprayers, corn pickers, and so on. These implements are difficult to mount on or remove from the tractor, and most of the time the tractor winds up just being bought and sold with this implement permanently attached. If the tractor you are considering has this hardware, seek out the advice and guidance of an agriculture extension agent, friend, local farmers, or other knowledgeable people to determine the implement's condition and the difficulty or consequences of removing it, if you plan to do so.

Special Considerations for Crawlers

Evaluating the condition of crawlers is tricky and calls for an experienced eye. If the crawler is to be used for production, I recommend hiring a heavy-equipment mechanic to accompany you to evaluate the machine. If the crawler is to be used in your collection and you feel you are mechanically inclined, you can evaluate the undercarriage adequately yourself by following the sidebar on crawler evaluation. These components, in addition to the steering mechanism, which may involve planetary gear sets or a differential, are very difficult for the novice or part-time hobbyist to inspect.

Price

Not much can be said about price within the confines of a book. I have seen tractors sell

While most collectors of antique tractors would love to own a steel wheel tractor, they are problematic to own. They are dangerous to load on trailers, destructive to asphalt and concrete, and often can't be entered into parades or tractor pulls. There is one solution, and that is to use parade tread. They are hard rubber coverings bolted to the steel wheels after the lugs are removed.

for $50,000, and I have seen them given away. About all I can tell you is that the price for the tractor you are considering should be somewhere between those two extremes! Prices vary considerably across the country. Don't let your cousin from Springfield tell you that you paid too much for your tractor since he bought his for $500 less. Unless you are willing to travel, prices outside your local area should not have too much influence on your decision.

Prices can vary considerably between seemingly similar models. Most people would see very little functional difference between a John Deere model 70 and the John Deere model 730, but I can assure you that the price difference will be tremendous (the 730 is more expensive). Also be sure to consider the cost and trouble of transporting the tractor home when negotiating the price. The price varies with the time of year, with most tractors selling cheaply at the end of the season, right before the holidays. Early spring and summer is the most expensive time to buy a tractor. The process of shopping for tractors will give you an idea of market price, and I encourage you to visit as many tractors as possible for this reason.

CHAPTER 4
Transporting Your Tractor

There are many reasons antique tractors need to be moved from one site to another. You may want to take it to shows, haul it to be repaired, take it to a remote work site, or simply get it home after you've purchased it. Whatever the reason, transporting your tractor will be a consideration sooner or later, and if you have never done this sort of thing before, it can be a perplexing obstacle. This chapter addresses the challenges you'll face and raises some safety concerns. First, we will cover some of your options and then cover the skills and equipment needed to haul the tractor yourself.

Don't Overlook Driving It!

The first option to consider is driving the tractor to your destination. Most states require only a slow-moving vehicle sign on the rear of the tractor and restrict access to certain roads. (Please check with your local highway patrol office before taking to the roads on your antique tractor.)

Driving the tractor is easier and cheaper than almost any other method and may be the only convenient option for many owners

Antique tractors are traditionally hauled on trailers, but they can be hauled in a lot of different ways. Here Brice Adams of Bloomington, Indiana, carries his Farmall Cub in an older rollback he purchased for hauling items in his collection.

who do not have a tow vehicle and trailer. It can also be a pleasant experience, a chance to travel the countryside at a slower pace than usual and to observe things about your home town that you ordinarily might not notice. The downside to driving the tractor is that it is unrealistic for long distances and can be more dangerous than hauling the tractor.

If you do decide to drive, consider your route carefully to avoid highways that are too dangerous, or illegal, for tractor traffic. You should also consider having a pace vehicle follow you. A pace vehicle will help prevent a rear-end collision with the tractor but may in certain circumstances create additional hazards. If the pace vehicle is large and obstructs the view of the tractor, for example, a motorist may try to pass the pace vehicle without realizing the tractor is ahead of it, creating the possibility for a dangerous accident. Take account, too, of how much fuel you'll need, whether driving alone or with a pace vehicle. Simply planning and thinking ahead will help to make the trip enjoyable.

Nothing could be easier than using a rollback and a winch to load an antique tractor. Generally speaking, you don't ride the tractor all the way up the ramp. You will just ride it up far enough to get it traveling in the direction you want and then let the winch take it the rest of the way.

If you can't call in a favor or barter for the haul, your next bet is to hire a rollback, or a trucker with a lowboy trailer. A rollback is a vehicle with a flat, movable bed. To load the tractor, the bed is tilted backward and slid out away from the cab and down to the ground so the rear edge of the flat bed is on the ground. The tractor is winched onto the bed and the bed is tilted up and slid forward so that the tractor rests flat on the back of the truck. It is then chained down for transport.

Hiring It Out

If driving the tractor is not viable, and you don't own a tow vehicle and trailer, then hiring out the work may be your only option. If you're lucky and have family or neighbors with hauling equipment, you might be able to get the job done for the cost of gas and a nice lunch. A local tractor club may also have a member with the time and expertise to make a short haul for you.

Antique Tractor Tip

Road Hog Hints
When driving a tractor on a road, a slow-moving vehicle sign is a smart addition that is often required by law. Flashing lights are a good idea as well. Magnetically mounted lights are available for restored tractors where these lights would detract from the tractor's appearance and value.

When winching or pulling a tractor, be aware the steering mechanisms or front axle may not be strong enough to handle the load and you could damage these items. This Farmall Cub's tie rods would be damaged by chaining to the front. Normally you would merely change directions and winch from the rear, but what if the rear of the tractor isn't accessible? Here we've set up how you might pull the tractor forward by hitching to the rear.

When shopping for a trailer, don't forget to evaluate the ramps and how they are stored. These side-stored ramps are handy and easy to get to and don't suffer from the ground clearance problems of ramps stored under the deck of the trailer. Long, heavy ramps are much easier to store and mount under the trailer.

Rollbacks don't have the capacity to move the very large antique tractors, such as a Cat D4 or a John Deere 830. The only other commercial alternative is a road tractor with a lowboy equipment trailer. This setup is required if your antique tractor weighs more than 10,000 pounds. These truckers have the equipment necessary to move any size antique tractor. The problem is finding one that will do the work in a reasonable time frame or for a reasonable price. Since the time to load and unload the tractor may not be insignificant (especially if the tractor is inoperable), the prices for professional truckers can be steep. Again, the local antique tractor clubs can provide the names of truckers who can help you out. Another option is to speak with the men and women who sell farm equipment, as they transport heavy farm equipment on a regular basis. They will either have their own equipment that they may be willing to hire out when they have a slow day, or they will give you the name of the transportation broker or trucking line they give their work to. As a last resort, look for transportation brokers in the Yellow Pages and get several quotes.

Hauling It Yourself: What You Need

If you've made the decision to haul it yourself, you've made a commitment to acquiring and using high-quality equipment and seeking advice and training.

There is more to transporting heavy equipment than just getting it onto a trailer and going. Successful hauling involves avoiding overloading your equipment, maintaining a safe operating environment for the people involved, avoiding damage to the hauling equipment and your tractor, and of course, bringing the tractor from point A to point B. If you do it right, whether the haul is a long-distance haul or a short one, a safe return and the sense of accomplishment and independence, not to mention the money you may possibly save (especially if your haul is a long one) are your rewards. I hope you will read this chapter closely, carefully plan your trip, seek the advice of trusted and experienced friends, and make safety your watchword. With patience, perseverance, and the right tools, you will be successful.

TOW VEHICLES

The first tow vehicle that comes to everyone's mind is a pickup. These are excellent tow vehicles that can be purchased in many different towing capacities. Whether your tractor is an 1,800-pound Farmall Cub or a 12,000-pound John Deere 830 with weights, there are pickups that can haul it. Typically, a good solid tow vehicle to haul small and medium-sized antique tractors is a full-sized pickup truck with a V-8 engine and a heavy-duty manual transmission or automatic transmission. The large tractors will require 1-ton pickups with full floating rear axles, duallie rear wheels, and a heavy-duty manual transmission. Here are some guidelines to follow when considering a pickup that will be used to haul tractors:

▶ Never use a compact pickup to haul anything but the very smallest of antique tractors (for example, a Farmall Cub).

▶ Always follow manufacturer towing recommendations.

▶ Never make the assumption that all pickups have generous towing capacities. Many V-6 pickups have very small towing capacities.

▶ Modern automatics handle towing very well. Older automatics are less likely to last when used extensively for towing.

▶ Manuals are great for towing, but you should be comfortable with a clutch.

▶ Manual transmission trucks often have their towing capacity de-rated by the truck's manufacturer. Check capacities when purchasing.

▶ Full-sized pickups should have a GCWR rating of 9,000–11,000 for small to medium-sized antique tractors. For large antique tractors, the pickup should have a GCWR of 15,000 pounds or more.

▶ Decide on a hitching system first (that is, receiver, gooseneck, and so on).

There are other tow vehicles available, all the way up to road tractors. Regardless of the vehicle you choose, make sure all your equipment is rated similarly and all the equipment can handle the load.

Towing Weight Considerations

The first thing you have to do is to estimate how much the tractor you will be hauling weighs. Owners' manuals typically list the tractor's weight, as do books that compile the Nebraska Tractor Tests. Be careful to exclude the weight of any ballast added to the tractor for the test (which unfortunately is sometimes not mentioned separately). If your tractor was not tested at Nebraska and you don't have an owner's manual (or the information is excluded), then contact the local area antique tractor clubs. There will be members in the club that could give you a fairly accurate estimate. In addition, you need to add the weight of any implements that may be attached to your tractor, the weights of any wheel weights added to the tractor, and any water-based ballast put into the rear tires.

Parker Yost shows safe form while operating a manual winch. He isn't standing and leaning over it, and he has the cable blanketed to prevent injury if the cable snaps. He also is using two hands.

The next step in estimating your hauling weight is estimating the weight of the trailer. This may be trickier than getting the tractor weight, as construction differs widely between trailer makes and models. Some manufacturers include weight on a manufacturer's

Antique Tractor Tip

Rent Rather than Buy
Consider renting a trailer whenever you need it. While many people find buying and owning a trailer is convenient, this isn't financially sound for most collectors and restorers, who may only need a trailer a few times a year.

When chaining antique tractors down to a trailer, a chain wrapped once or twice around a hitch is still able to move from side to side as the hitch bar slides through the chain wraps. While the range of movement is typically slight, it might be enough to loosen chains, which in turn may cause binders to fall off, and then there is a very real danger as you travel. To prevent sideways movement of the tractor as you chain down the rear, center wrap on the hitch ball as you see here, use a clevis, or include the side arms of the hitch when routing the chain.

ID plate on the trailer. You may be able to contact the manufacturer of the trailer, but if that is not possible, the only safe way to estimate its weight is to take the trailer to a weigh station on the highway or use the scales at a nearby truck stop or feed store. If you need to do this, take the trailer to be weighed using the tow vehicle you will use to haul the tractor. That way you can weigh the tow vehicle and the trailer separately and have a very accurate idea of the weight of the hauling equipment you will be using. Once you know the weights of the equipment involved, it is time to match them up to manufacturers' recommendations.

GVWR and GCWR

There are three primary tow vehicle manufacturer weight ratings you need to be aware of when you haul a tractor. The first two are the tow vehicle's gross combined weight rating (GCWR) and its gross vehicle weight rating (GVWR). The GCWR tells you how much the vehicle, and everything it is pulling, can weigh. If the GCWR is 10,000 pounds, then the trailer, tractor, truck, occupants, and their luggage and tools combined cannot weigh more than this or the vehicle will be dangerous to operate. The GVWR is the total weight that can be placed in the tow vehicle. Be sure to estimate all the occupants and the weight of any cargo. The third is the tow vehicle's tongue weight rating. This is the total amount of weight that the trailer may place on the hitch of the tow vehicle. Be sure to include this tongue weight in your GVWR calculations for the tow vehicle.

The trailer also has a gross vehicle weight rating, reflecting the maximum load you can place on the trailer. This rating includes the weight of the trailer, so a trailer with a GVWR of 5,000 pounds cannot carry a 5,000-pound tractor. It can carry 5,000 pounds minus the weight of the trailer. This rating may be expressed on a per-axle basis, since many manufacturers make trailers in many different axle configurations. The trailer will also have a hitch rating, reflecting the amount of weight that you may pull with the hitch. The hitch rating applies to the total weight of the trailer and its contents combined. For example, the 5,000-pound hitch may pull a 3,000-pound trailer with a 2,000-pound tractor, or it can pull a 5,000-pound trailer empty. The hitch will usually also have its own tongue weight rating, though typically it is much higher than that of the tow vehicle and therefore not a limiting factor. The hitch may also have a second, higher weight rating if you use a load- or weight-distributing hitch head in the hitch receiver.

Hitches

There are several different kinds of common hitches. The first, and lightest duty, are the bumper hitches. The requisite hitch ball or receiver mounts directly to the bumper and the trailer attaches to this. Bumper hitches do not have

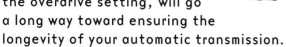

Antique Tractor Tip

Cool Your Transmission
If your tow vehicle has an automatic transmission, be sure to add a transmission cooler to it. These devices, in addition to limited use of the overdrive setting, will go a long way toward ensuring the longevity of your automatic transmission.

the weight rating necessary to tow tractors. They may, however, have the weight rating necessary to pull a trailer that is carrying a small implement or tools. The next and most common type of hitch is the frame-mounted receiver style hitch. This popular, heavy-duty hitch is bolted or welded to the tow vehicle's frame and is available in weight ratings that go much higher than many tow vehicles can handle. The limitation to this type of hitch is tongue weight—the weight the loaded trailer exerts on the hitch. While the hitch may be able pull a 7,500-pound trailer, the trailer may put too much tongue weight on the tow vehicle, so the load must be carefully positioned or a different style of hitch will have to be used.

The three common extremely heavy-duty, high-weight-rating hitches are the fifth-wheel, gooseneck, and pintle (or Ring and "D") hitches. The fifth-wheel hitch and the gooseneck hitch both are placed in the bed of a pickup truck. The fifth-wheel hitch is a small version of the same type of hitch road tractors use. It is usually mounted to the frame of the truck on the bed between the wheelwells, and the hitch surface itself sits slightly higher than the tops of the wheelwells. Large horse trailers are often pulled with a fifth-wheel hitch. Gooseneck hitches sit flat in the bed of a pickup truck, and hold the beak, or the end, of the gooseneck portion of the trailer. The main advantage of these hitches is that they place the tongue weight of

the trailer above the rear axle of the tow vehicle. First, this enables the tow vehicle to handle higher tongue weights. These heavy-duty hitches have "kingpin" weight ratings instead of "tongue" weight ratings, but the idea is the same. Pintle hitches are heavy-duty hitches used to carry very heavy loads. They are primarily used on tow vehicles that cannot have a bed type of hitch, like a dump truck. The hitch on the truck is a very heavy-duty hook that opens to hold a ring on the tongue of the trailer. The hook is stationary or may close and catch to form a "D" on the hitch. The next time you are driving along and see a large piece of construction equipment like a backhoe being pulled by a dump truck, look at the hitch. Odds are it will be a pintle hitch.

Lighting, Braking, and State Regulations

There are very specific traffic laws that address pulling a trailer behind a tow vehicle. The laws are fairly uniform from state to state and are modeled after the federal highway regulations. There are differences, however, that you must be aware of. In some states, brakes are required on at least one axle of all trailers, regardless of weight. Some states, such as North Carolina, require brakes on one axle only if the complete weight of the trailer and load exceeds 4,000 pounds. In some states there are exceptions and variances for agricultural loads, so you must be aware of those too. In addition, almost all states require brakes on all axles over a certain weight.

When purchasing a trailer, you'd be hard pressed to find one that is a better configuration than this. It has dual axles, electric brakes, a wooden deck, a center-mounted jack well away from the tongue, and side rails. It is long enough for any small to mid-sized antique tractor. While a few other options, such as a toolbox or beaver tail, would be nice, this trailer would do the job.

Virtually all states require that the trailer have rear lights that behave exactly as the lights on the rear of your tow vehicle. Some states require additional running lights for longer or enclosed trailers. Many complete road atlases contain summary information on trailering laws from state to state.

Toolboxes are handy additions to trailers. A lockable front-mounted tool box like this one is just the right size for stowing chains, binders, and tools.

Antique Tractor Tip

Fifth-Wheel Wisdom
If you plan on one day owning horses or using recreational vehicles, consider installing a fifth-wheel hitch, as they are commonly used for these types of trailers.

Some additional trailer equipment is considered necessary and prudent, even if not required by law. Using side running lights on a trailer is an excellent idea. In addition, older towing vehicles with drum-only braking systems should only tow trailers with brakes. Safety chains, which hold the trailer to the tow vehicle if the hitch fails, are a requirement in every state. Break-away trailer brakes are an excellent idea as well. These brakes are wired to a battery on the trailer and are automatically engaged to stop the trailer if it becomes disconnected from the tow vehicle. If you are going to buy a trailer with brakes, I recommend also buying the break-away kit even if they aren't required for your particular circumstance. Be sure you have an adequate spare tire for the trailer and a jack powerful enough to lift the trailer when loaded.

LOADING THE TRAILER AND SECURING THE LOAD

Properly loading and securing the tractor onto the trailer are extremely important steps and can make the difference between a safe hauling effort and a very dangerous one. Loading the tractor is hazardous for many reasons. This is because loading ramps on trailers are very steep, and tractors are unstable by nature on inclines. Trailer ramps are also narrow, allowing a tractor to fall off the side of the ramp if the operator isn't careful. Wet, painted, or metal trailer surfaces can be very slippery and I have seen a number of tractors very nearly slide off trailers because of these factors

while being loaded. Tractors can also flip or overturn during the loading process. A safe loading plan is therefore essential any time you are going to haul your tractor.

Setting up for loading will come first. Move the tractor, if possible, to a level, firm area. The act of driving the tractor up onto the trailer will tend to raise the rear of the tow vehicle. The rear of the tow vehicle is usually the axle with the parking brakes, so if the rear raises enough to lose traction, this will allow the trailer to shift and move and cause problems. I like to block the trailer by laying large pieces of wood on the ground under the back frame members. This keeps the rear of the trailer from dropping very far. Ramps with cleats, rough surfaces, or slats are necessary to keep the tractor from sliding down the ramps. Wood ramps are satisfactory only for the lightest tractors and in dry conditions. Be sure to use a middle ramp that anchors to the trailer if your tractor has a narrow front wheel arrangement.

I prefer to winch the tractor up onto the trailer if the driving conditions are anything less than good. If you must operate it, try to have helpers to watch the behavior of the ramps and the trailer and tow vehicle so you can concentrate more fully on driving the tractor. Consider backing the tractor up onto the trailer rather than driving straight on if the ramps are particularly steep, or the trailer and tractor are on any kind of incline (loading on an incline increases the dangers, so avoid it if possible). Once all wheels are on the trailer safely, stop it a few feet before you think you should. The operator should stay on the tractor and have an assistant pull the tractor forward the last few feet while the operator uses the brakes, as necessary, to keep the tractor under control. This allows more precision in trying to balance the load on the trailer.

Your goal is to place the tractor so that 60 percent of the weight is in front of the pivot line of the trailer. This pivot line is the axle of a single-axle trailer, it is between the two axles of a double axle trailer, and so on. It is better to err on the side of having too much in front of the pivot line (for example, 65 percent) rather than having too much weight at the back of the trailer. There is no easy way to determine when you have the load properly distributed other than by eye and by feel. Also judge how much tongue weight is on the tow vehicle and be sure you are not exceeding it. Verification of tongue weight can be accomplished at scales by weighing the truck with the trailer attached but with the trailer off the scales. Then pull off the scales, detach the trailer, and weigh the truck again. The difference is the tongue weight.

Binding down the load is very important and should be done thoroughly. Use stout chains capable of handling the load; at the very minimum, use 3/8-inch cold proof steel chain. Use a larger size for large tractors, crawlers, and so on. Tighten the chain using chain binders designed for the size of chain you are using. Ratcheting type chain binders are the handiest and easiest to tighten,

but lever type binders work well also. When attaching the chain, remember that binding the load from side to side is less important than securely binding the load at the front and back. The strongest force the tractor will encounter is that of the braking effort of the tow vehicle. Therefore, bind down the back of the tractor very well. Chain the rear at two places to the frame of the trailer or to any D rings intended for the purpose. Then bind down the front of the tractor very well, again at two places to compensate for acceleration forces. If you were able to bind the front and the rear at two places and you were able to attach the chain to the trailers near the corners, then a chain at the sides of the tractor are not necessary. I often do chain at the sides for added insurance in case one of the other chains gives way, but the tractor will not flip over if the front and rear chains remain properly secured.

There are a few tips and tricks for unloading and loading a tractor. I like to back the trailer up to a small hill so the back of the trailer touches the ground. Then there is no need for ramps and you can load the tractor by driving straight onto it. If you are unloading and you had to winch the tractor up onto the trailer because the tractor was not working, then the easiest way to unload it is to simply roll the tractor off the trailer, allowing it to roll down the ramps if at all possible. There are limitations to this procedure. If the steering system of the tractor is shot,

This trailer represents what would be considered a dream trailer for hauling antique farm equipment. It is a gooseneck configuration, which means that it pulls behind the tow vehicle more easily and has more stability. It has a spacious toolbox mounted on the gooseneck frame within easy access and is out of the way of the equipment being loaded. It also includes a winch for loading equipment. The rear of the trailer has a dropped end known as a beaver tail for less extreme loading and unloading angles. This trailer also has three lay-down ramps. The third middle ramp is necessary for narrow front tractors.

Antique Tractor Tip

Consider Custom Trailers
At most trailer manufacturing and retail centers, a custom trailer can be purchased for nearly the same money (option for option) as a stock trailer. Adding things like a winch bar and a third ramp for your narrow front tractor are easily ordered, and waiting times for the trailer are usually reasonable.

wheels may leave the ramps early and drop to the ground. Also, the tractor may roll for quite a ways after it leaves the ramp, so make sure people, pets, and other vehicles are well away from the unloading area. If the tractor had to be winched onto the trailer because of a frozen transmission or frozen tracks, then winching it back off is the only option.

The only other consideration that needs to be addressed is actually transporting the tractor. Before starting off, look over the tractor one more time, making sure you have removed any loose items like seat cushions, tools left on the tractor, and so on. Once out on the road, there are several very important driving tips to keep in mind. During the first couple of miles, drive slowly and make sure there are no problems cropping up with the trailer, the load, or how you have secured it. After you have been traveling for several miles, stop and retighten your binding chains. Remember that you need to make wide turns when you are towing a trailer.

Leave plenty of extra braking distance between you and the vehicle in front of you. Be wary of cross winds, and pay attention for any sign that the trailer is going to try to start swaying back and forth. If it does, touch the brake. If the trailer has brakes, this strategy is very effective and the trailer should settle down quickly. If not, then touching the brake momentarily helps, but then you will need to keep your foot off the accelerator for a while until the trailer settles down. Swaying happens during cross winds and it can also be caused if the load on the trailer isn't distributed correctly. If you get any swaying that does not seem to be from a cross wind, stop and move the load forward to increase tongue weight. Be patient during transport and remember that you need more time and distance to brake, turn, and accelerate than usual.

Learning to back up a trailer requires patience and repetition, and I suggest practicing with an empty trailer in a large parking lot many times before your first

trip. When backing up with a trailer, keep two things in mind: The trailer reacts quickly to any change in the steering wheel, and the trailer moves in the opposite direction of the rear of the tow vehicle. One trick you can use when backing the trailer is to place your hand on the bottom of the steering wheel and turn the bottom of the wheel in the direction you want the trailer to move. For example, if you want the rear of the trailer to move to the right, move the bottom of the steering wheel to the right (counter-clockwise). It's that simple. Keep in mind that slow speeds are best and small changes at the steering wheel can create large changes in the direction of the trailer. Be careful and remember that practice makes perfect!

Regardless of how you decide to move your tractor from point A to point B, transporting it will eventually become a necessity and being safe and successful are the goals. Hauling your tractor yourself can be rewarding and fun. Common sense ideas about safety, using quality equipment that is in good repair, and using the equipment within the limits it was designed for will go far toward making sure you survive to be able to haul tractors for many more years. Hiring the work of hauling your tractor is also an excellent idea, especially if the tractor is particularly heavy or you are short of time or equipment. Either way, understanding your options and the skills involved will help you make the right decision for your circumstances.

Starting and Operating Your Tractor

Once you have a tractor home, sitting in the yard, how do you go about putting it to work? The first place to start is, well, getting it started and driving off.

While operating antique tractors is quite simple, you must be familiar with the machine before you go plugging away. Before you load up the machine of your dreams and head home, grill the owner on every aspect of the tractor. What does each lever do? How do you get it started? What are its starting idiosyncrasies?

Don't be afraid to ask lots of questions and take notes if need be—the previous owner is likely to be your best source for specific information.

Antique Tractor Anatomy

If the array of knobs, levers, gizmos, and gauges were not enough to overwhelm the novice antique tractor owner, the vast differences of the same controls between models will. Clutches for many John Deere tractors do not behave, nor do they engage, the same way that a Farmall does nor even the way a New Generation John Deere does. Mechanisms for activating the starter vary, proper oil pressure readings vary considerably,

Hand-starting tractors were common prior to WWII, and here Francis Robinson is getting ready to start his Minneapolis-Moline. When hand starting, remember to engage the starter crank with the handle down and turn the engine by pulling up on the crank. If you are pushing down, you increase the likelihood of injury should the engine kick-back.

Many antique tractors have an electric start and a hand start as a back up, as this Case does. When looking to buy any tractor that has a hand start, make sure the crank handle is present. Many crank handles, such as this one, are cast-steel or -iron and can be expensive to find or reproduce.

and the feel of the various steering mechanisms between tractors differ quite a bit. What I will present here is a bit of an anatomy lesson about antique tractors that I hope will demystify the bewildering forest of levers, knobs, and switches that confront you.

Throttles and Choke

Throttles are typically hand levers and are almost always situated around the steering column. These come in all shapes and sizes. In addition, tractors have a governor that constantly maintains engine speed based on the throttle setting. The throttle and the governor work in unison to maintain a certain rpm level from the engine. The throttle does not directly

affect the carburetor as automobile throttle pedals do.

The choke for antique tractors is meant to enrich the air-fuel mixture drawn into the carburetor (giving a higher proportion of gasoline relative to air) for cold engine starting. Virtually all chokes work by restricting the air intake passage of the carburetor and activate via a cable or rod. The end of this cable should terminate somewhere near the dash. Often this cable is rusted to the point of nonoperation or missing, with a length of mechanic's wire in its place. This works fine in a pinch and you can fashion your own should yours be missing or broken.

Ignition

The ignition switch can vary quite a bit from one tractor to the next. Some, such as my Massey-Harris Pacer, the Ford N series, and many others operate with a keyed switch. The older tractors tend to have some simple type of knob that you turn much like a key to the "on" position. How many positions the switch has depends on whether the tractor has a magneto ignition or a battery ignition, complete with lights and a charging system. There will be at least two positions, on and off. A third may exist for lights, and there may be additional positions to set the level at which the charging system will operate. Since ignition switches almost never serve as the starter switch, there will not be a "start" position, as on cars.

Kill Switches

In the event of an emergency, it is handy to have a strategically placed kill switch to stop the engine immediately. This switch is designed to be easily operated and is placed in a spot that is easily reached. The ignition switch, which tends to be placed lower down on the dash, is difficult to reach quickly and may not be as intuitive an action as reaching out for a kill switch. Kill switches were almost never included as standard equipment from the factory, so these are aftermarket retrofits. Before relying on this sense of security, test it and make sure it works! By the way, some antique tractor shows and pulls require you to have a kill switch installed on your antique tractor.

Lights and Gauges

Lights were available as an option on tractors as early as the teens of the twentieth century. Typically, the earlier tractors (if they had lights) had one or two forward-facing lights. Later tractors (1930s on) also included rearward-facing lights, especially if the tractor was equipped with a PTO or a belt pulley in the rear. Most electrical systems were 6 volts, though many antique tractors have had their electrical systems upgraded to the more modern 12-volt negative ground system; be sure to check your electrical system voltage if you have any doubts. Sealed beam headlights became common around World War II or shortly thereafter. Modern replacements are often available for these. Cigarette lighters became popular after World War II and have remained so.

The gauges on antique tractors were sparse and simple. The oil pressure gauges are similar to modern gauges, but typically register much lower oil pressures than modern gauges. The coolant temperature gauges are uncommon in earlier tractors and were associated with nongas engines such as distillate engines. Gas tractor engines typically run cooler than modern equipment, and normally had only 140 to 180 degree thermostats installed at the factory. Kerosene and distillate tractors were designed to run hotter. Ammeter gauges show the health of the charging system if the tractor is equipped with a generator and battery. The gauges should read in the charging area unless the lights are on and the engine is not at full throttle. Some of the younger antique tractors have tachometers, which are handy for PTO work, and hour meters, which help the owner-operator keep track of the usage of the machine, production per hour, and so on.

CLUTCHES

Clutches actually come in many forms on antique tractors. Probably one of the most endearing is the hand clutch, operated by a hand lever, of course, rather than a pedal. Many older tractors, including the Massey-Harris 55 and John Deere models prior to the mid-1950s, had hand clutches. They were also used on orchard models from several manufacturers. Another clutch variant, found on many younger antique tractors, is the two-stage foot clutch. With this clutch, when the clutch pedal is depressed about halfway the drivetrain becomes disengaged but the accessories driven from the engine, such as

the hydraulic pump and lift and the PTO, are left engaged. This allows you to continue operating these items without cessation while stopping your forward progress, which is handy with implements such as a hay baler.

FOOT PEDALS

Foot pedals tend to be straightforward but differ from cars in that tractors have two brake pedals, or occasionally three. This is so the right rear brake and the left rear brake can be operated independently. This helps in steering the tractor in the field and is typically reserved for heavy tillage operations. The brakes can also be operated together by using a device that locks the two pedals together, by using the third pedal, where available, or because the placement of the pedals allows the operator to hit both pedals at once. The brakes are operated simultaneously while traveling on the road, performing nontillage chores, and in other circumstances where even braking is desired. In addition, there should be some provision to lock the brakes themselves for parking the tractor. The two brake pedals must be locked together before setting the parking brake (except on three-pedal models).

IMPLEMENT LIFTS, PTO, AND BELT PULLEY

One of the things that distinguishes a tractor from almost everything else is its ability to pick up and power implements. The lifting arrangements varied quite a bit over the early years, both across companies and across models. Harry

Many antique implements are driven by belts, and most antique tractors have belt pulleys for driving them. Here the author's 1952 Oliver 77 diesel is belted up and ready to work. To maintain the right tension, the tractor is backed up and a wood block is placed under the rear wheels.

Ferguson then designed and built the first functional three-point hitch. This hitch not only allowed the operator to pull and lift an implement, but also allowed for the automatic height, or "draft" control of the implement. For example, plows require their inclination, position, and depth relative to the surface of the ground to be fairly precisely situated to operate efficiently and well. The Ferguson three-point hitch solved this by sensing the load at the top, or the third point, of the hitch. The hydraulics then pushed or pulled the lower link arms in response, altering the geometry of the plow and hitch. Harry Ferguson's first designs and demonstrations in England were successful, and his subsequent business deals with Ford have become legendary. Refer to the Ford History section of this book for more information.

PTOs, or Power Take-Off shafts, were early inventions in the tractor industry. Their size, speed and rotations were standardized

The controls on an antique tractor are fairly straightforward, but they are a bit different than folks without experience on tractors might expect. The primary difference is usually a lack of an ignition key. Here, the small silver lever seen just above the gear shift knob operates the ignition and lights. All antique tractors have a manual choke, and the handle of the silver lever is almost pointing directly to the black knob that operates the choke. The silver button below the black knob is the starter button. That is what you push to start the tractor. Rounding out the controls are the throttle lever to adjust engine speed (mounted just under the steering wheel), an implement depth adjustment lever (to the right on the gas tank), and the gauges, which operate and read as anyone with experience with an automobile might expect.

in the 1930s. The PTO was adopted nearly completely by all manufacturers for all their models, with some notable exceptions, such as the Farmall Cub. The standard shaft is 1 3/8 inches thick with six splines and rotates clockwise, turning at a "no-load" rate of 540 rpms with the engine running at the rated speed. There should be a clear hole for pinning the output shaft to the PTO shaft. If the PTO shaft is not "live," then there might be an "overrunning" clutch attached to it. Overrunning clutches prevent the driven implement, such as a bush hog, from acting as a flywheel and continuing the forward movement of the tractor even though the clutch has been depressed.

Belt pulleys differ quite a bit and may show up anywhere on a tractor. The smaller tractors seem, as a rule, to have their belt pulley installed at the rear of the tractor, while the larger tractors have a belt pulley situated mid-vehicle, usually right in front of the operator on the right-hand side. They are located so the belt will miss the front wheel and axle; some old gas and steam traction engines have front axles that completely pivot out of the way for this reason. They all generally have, when viewed with the rotation axis running left to right, a concave surface, which helps to keep the belt from slipping off during use.

HAND LEVERS

So many levers, so little time! Many specialized implements will have mounting and operational designs peculiar to your tractor. A primary example would be manually

operated cultivators. These were operated with one or more hand levers that were spring loaded and operated a rockshaft to which the cultivator was mounted. The rockshaft and cultivators may be long gone but the levers for some reason seem to stay on the tractor. There will be many other levers that engage the PTO, operate the hydraulic lift, or even operate the clutch. Refer to your operator's manual for specific information.

Systems Check and Fueling

Preoperational systems checks for tractors are very similar to those for a car. Things like general condition, tire inflation, and presence of children, pets, and animals should be noted when approached. There are a few additional items to note with antique tractors. The field conditions should be noted, as ground that was firm when the machine was parked could be slick after a rain, presenting interesting challenges. Implement position must be noted. I once nearly had a serious accident when I neglected to check an implement before using the tractor. I parked the tractor with the implement raised, but because of a weak hydraulic system, the implement had drifted to the ground. I returned to the tractor assuming the implement was still raised, and I fired up the engine and let out the clutch to proceed. While the implement was not imbedded in the ground, there was a root directly in front of it. Tractors, if they cannot move forward, will rearward flip because the rear wheels will drive the body up and over. Attached as my tractor was to an immovable object, the front end began to come off the ground. Fortunately, before it got more than a few inches in the air, the root broke, leaving me safe and sound but completely scared. The front end started to rise faster than I ever thought imaginable. It all would have been avoided with a quick glance at the implement before proceeding.

Different antique tractors have different fuel needs, and using the proper fuel is of course vitally important. Which fuel your tractor needs must be researched if you don't know. Don't assume that gas is the proper fuel! While many engines designed to run on other fuels will run on gas, there are some cooling system issues that will need to be addressed to do so. Local farmers and mechanics can help identify the proper fuel for your tractor, but again, refer to your operator's manual. It will inform you of which fuel to use and will give important precautionary measures.

Leaded versus Unleaded Fuels

If gasoline is the fuel that you will be using, then you must decide whether you should be using a lead additive. Many years ago, lead was added to gasoline to increase its octane rating, but a side benefit was the lubrication of valve seats and valves during operation. When lead was removed from gasoline, this lubricating property was lost. Exhaust valve seats will erode without leaded fuel if two conditions are met: one, the valve seats are not made of hardened material, and two, the

engine is run hard for long periods of time. Fortunately, one condition or the other is usually not met for most antique tractors. Tractors that see regular use are usually only worked for light chores once or twice a week, for a few hours at a time. If the tractor is worked much harder than that, its engine may well have been rebuilt and fitted with hardened valve seats.

When looking at a tractor, ask how much it is used, what its fuel requirements are, and whether the engine valves have hardened seats. Fueling an antique tractor is straightforward; however, there are a few nuances. The tanks in antique tractors are often difficult to reach because they are inconveniently placed or partially obstructed by an implement, such as a front-end loader. Spilling fuel while filling the tractor can be dangerous, as often the battery is close by. I have two acquaintances who have suffered serious burns from fueling tractors, so please exercise extreme caution. Never fuel in a barn or in any building with a lot of easily ignited material. Another thing to remember is that most older enamel paints are not resistant to gasoline, so wipe up after a spill if you want the paint to stay in good shape.

Although gas caps were often vented, sometimes they weren't and sometimes the vents became plugged up over time. The pressure release when you open a tank without a vented cap can be dangerous if the tractor is hot, so please allow the tractor to cool a bit and ease the cap off. I prefer to fill my tractors with small, portable four-gallon gas cans, as the large cans seem to spill more easily.

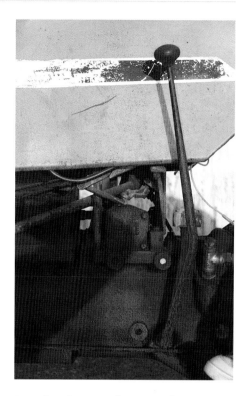

Many tractor manufacturers began offering systems that were, in effect, a high-low transmission that allowed the operator to shift down for additional torque, albeit at a slower speed. The tractor was operated primarily in high, but when difficult ground was encountered where extra pulling power was needed or a thick place in the windrow was encountered, the operator could shift on the fly to low and give the tractor the extra lugging capacity at the expense of speed. As soon as the difficult patch passed, the operator would return the unit to high. This lever is a high-low shifter on a Case tractor, but many makes had them, including Minneapolis-Moline and Farmall with its Torque-Amplifier. All high-low units were in front of the transmission.

If you have a tractor with two gas tanks, odds are it was designed to be started on gasoline, then operated on a lower grade of fuel. The other possibility is that the tractor has a large diesel motor and the smaller second tank is for the fuel to run the small gasoline starting, or "pony," motor. Distillate engines are impossible to start and usually have

Implements are attached and driven by a variety of endless designs if they aren't your standard three-point/PTO implement. Mid-mount implements are the most convoluted and usually incorporate a series of brackets, arms, plates, shafts, and belts to attach and drive the implement. Here the upper belt cover was removed for the photo, but always operate these implements with covers attached.

an auxiliary starting tank that you put gas in. You start the engine on this gas, and after engine warm-up turn a valve in the fuel supply line to turn off input from the gasoline starting tank and open the input line from the larger distillate tank. Distillate tractors require a very warm engine to run efficiently, so radiator shutters, curtains, or coolant return valves were often added to restrict the engine's cooling capacity. If you decide to run your tractor only on gas, which will work just fine, remember to leave these components in the "open" or "gas" position so the engine will adequately cool.

Starting Antique Tractors

For those not accustomed to older equipment, the procedures required to start an antique tractor can range from the obvious, to the not-so-obvious, to the nearly ridiculous (certain models of Field Marshalls

start with explosive shells similar to shotgun shells). I have owned several antique tractors and have operated dozens of others, and none of them started quite the same way. Some started with a hand crank, others had an electric starting motor that was operated with a push button, some a pushrod, some a key, and even a knife switch in one homemade contraption I looked at one time. Often the tractors start on one fuel, such as gas, and run on another, such as distillate, once the engine is brought up to operating temperature. The variation of starting procedures seems endless and is one of the charming aspects of antique tractors.

Manually starting an engine that starts on gas generally means either turning a hand crank that engages the crankshaft pulley nut, or spinning a flywheel by hand. In either case it is imperative that you set the parking brake or chock the wheels. During these circumstances, you are very close to the tractor but not in direct control. If the tractor is left in gear and lurches, or left in neutral and rolls, it can create a dangerous situation. Immobilize the tractor first before starting it, double and triple checking to make sure the transmission is in neutral. Use the following guidelines when starting any gasoline engine antique tractor.

1. Immobilize the tractor, using chocks and the parking brakes.

2. Do complete preoperational checks and fueling.

3. Check the radiator fluid level and oil level.

4. Put the transmission in neutral.

5. Turn the fuel system on. Usually, a valve in the sediment bowl or the fuel line needs to be opened to begin the flow of fuel to the carburetor. Liquid propane (LP) fueled tractors have main valves at the tanks that must be turned on.

6. Remove nonstationary weather caps from the muffler.

7. Open compression petcocks or other compression-release devices to aid in starting. (Not all tractors have these devices.)

8. Choke the carburetor if the engine is cold or the weather is extremely cold.

9. Turn the ignition on.

10. If possible, also disengage the clutch (many older tractors have a clutch stop that keeps the clutch disengaged while the operator is off the tractor).

11. Start the tractor. The following paragraphs outline this step further.

Once the engine starts, close any compression petcocks that were opened to assist starting, inspect the operation and condition of the engine and gauges, making sure there are no fuel leaks, and so on, and prepare the tractor for use.

There are several different ways to start a hand-crank tractor, but only one way that minimizes your risks. Begin by engaging the hand crank on the crankshaft pulley nut and turn the engine gently and slowly until the engine's compression starts generating resistance. Then remove and re-engage the hand crank so the handle of the crank is near the very bottom of its revolution, that is, at the 5:00–6:00 position. Stand in front of the hand crank facing the tractor. Take one step to the left and then a half step forward and then turn to face the hand crank. To crank the engine you will grip the handle of the crank with your strong hand, its palm facing away from you and your strong leg slightly in front of the other leg. Don't wrap your thumb around the hand-crank handle; place it beside your index finger. To crank you will pull upward as smartly as possible, using leg strength to begin the stroke and arm strength to finish it.

You will only actually crank about 180 degrees, always pulling up and never pushing down on the hand crank. Never crank around in a full circle, as the engine can sometimes kick back. If it does, the stance and procedure just described will keep you out of harm's way. To continue the cranking motion all the way around puts your body and face in the way of the crank handle. (If you think I'm kidding about kickback, ask any old-timer how many broken forearms he knew about that resulted from cranking Model T engines.) Of course, you may need to repeat this procedure a number of times. If the engine doesn't start, refer to the troubleshooting section of this book.

On tractors without hand cranks but with manual starting, you will need to spin up the engine using the flywheel. This is the method for starting all two-cylinder John Deere manual start tractors. After following the prestart checklist mentioned earlier, simply grasp the flywheel, turning it counterclockwise until resistance in the engine is felt. Then turn as smartly and quickly as you can.

If the tractor has electric start, you are in luck. A charged battery is all you need to start the engine. To determine exactly how to engage the starter, find the starter and determine if there is a heavy-duty mechanical switch on top, or a canister-type magnetic solenoid switch. The former will have some type of mechanical linkage attached to it. This linkage allows you to remotely switch on the circuit from the battery and engage the starter. The latter is a magnetic switch that is electrically operated and is identical to the starter circuits of cars. In cars, the key activates the starter, while on tractors it is often a push-button and the key just operates the ignition, charging, and lighting. To start a tractor with electric starting, simply follow the same directions, but remain on the tractor and substitute activating the starter for spinning the crank or flywheel.

If your tractor is running diesel, you may have the simple task of engaging the electric start, or you may have the more involved procedure of firing up a pony motor and starting the engine with that. The only revision to the above starting procedure you would make for a diesel engine starting on an electric starter would be the fact that there is no ignition to turn on and

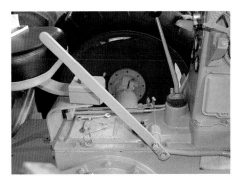

The lever you see here is the hand clutch and is something you'll never see on a modern tractor. This Minneapolis-Moline, all older John Deere tractors with the horizontal two-cylinder engines, and many others have them. The clutches themselves operate on mechanical principles nearly identical to foot clutches, it's just that there is a hand lever instead of a foot pedal to operate the clutch. It takes a bit getting used to, so be sure to practice in a open, flat, and safe area the first time you operate one.

there may be intake manifold glow plugs to turn on. These glow plugs warm the intake air stream and help to make the engine easier to start (do not use starting fluid in a diesel engine with manifold glow plugs!). By the way, diesel engines have no spark plugs and there is no need for ignition current. Simply turning on the fuel source readies the engine for starting. Likewise, shutting off the fuel source is the only way to stop a diesel engine. Fuel shut-off is accomplished through a valve in the fuel pump that can be remotely operated by the operator.

Starting a diesel tractor with a pony motor is more complicated. First you should follow the above gas engine starting guidelines, but you should apply them to the pony motor and not the diesel engine. This little motor will be the motor that turns and starts the larger diesel engine. These usually have their own electric

Some tractors had hand brakes instead of foot brakes. This Massey-Harris 55 exhibits one—it is the large lever to the right of the fender. They are differential brakes just like foot brakes on all antique tractors. That means it is possible to apply brakes to one side at a time and help steer the tractor with the brakes.

starts and will start and run the same as any small gas engine. The pony motor and the diesel engine may share coolant, or the exhaust stream from the pony may be routed around the intake air manifold of the diesel so the warmth generated by the pony motor makes the diesel easier to start. You will want to leave the pony running for anywhere from 20 seconds to a few minutes depending on several factors. These factors are the ambient air temperature, whether or not heat generation from the pony is used to prewarm the diesel, and how difficult your diesel is to start. Under most circumstances, the diesel can be spun up just as soon as the pony has had a chance to warm up on its own—about 15 to 20 seconds. After turning on

the fuel source for the diesel, you then let out a clutch that engages the pony motor to the diesel motor (usually via the flywheel) and starts the diesel spinning. The diesel should then fire up after several revolutions. Disengage the pony clutch, kill the ignition to the pony motor, and your diesel should be up and running on its own. Again, if you encounter problems, please refer to the troubleshooting section of this book.

There are a couple of other starting variations on diesel engines. IH used a diesel engine that started on gas. These engines have a separate combustion chamber with a spark plug as part of the diesel combustion chamber. This allows the engine to start on gas, but an ingenious one-step mechanism allows complete cut-over to diesel by closing the chamber, shutting the carburetor, and enabling the diesel pump.

Engines of this age period tend to benefit from a short warming-up period after starting. I like to idle the tractor's engine for a few minutes before heading out to do chores with it. The trip to the work area finalizes warm-up. If the work is particularly close at hand, and the tractor is already set up and ready for work, I will allow it to idle for another few moments before I begin. I like the engine (and hydraulic fluid) to be at, or near, operating temperature when the real work begins. Likewise, I allow a cool-down period at the end of the day. I let the tractor idle in its final parking spot for a moment or two. I have no evidence to support that this does any good, but I have heard this advice over and over again all my life and I am sure it does no harm.

STARTING FLUID

Engines difficult to start sometimes can benefit from starting aids like starting fluid, also known as "ether." (By the way, WD-40 works in place of starting fluid.) I have never found a well-kept, well-maintained engine in solid mechanical shape that required starting ether to start unless conditions were extremely cold. The fact is, though, not all engines are well kept, maintained, or solid and occasionally it is extremely cold, so a starting aid is needed. To do this, spray a small amount into the air intake of the engine. Never use large amounts, only a small whiff. If the engine will not start within a few applications, the underlying cause of the trouble should be addressed before you continue to use ether. Starting fluid used improperly, too liberally, or in conjunction with a malfunctioning ignition system can damage your engine. Remember, too, that starting ether should never be used in a diesel engine with glow plugs!

Operating the Tractor

One of the first things you will notice about antique tractors is the impressive amount of torque they generate. It is very clear from the moment you hop onto an antique tractor that very little will stop this machine. While you may not go more than 10 miles per hour, you could pull a house behind you at the same time. This is the heart of tractor design and the fundamental difference between them and other machines in our lives. They are intended to pull heavy, difficult, dynamic loads in a productive and safe manner.

Now that our tractor is started and running, it's time to move out to become acquainted with it. The transmission is not very different from a car's—typically three to five gears and a clutch. Unlike cars, however, tractors are designed to start out in almost any gear. It is not necessary to start out in first, then move through the other gears. If second gear is the gear you wish to use, you simply start out in that gear. In fact, in tractors made before the advent of helically cut gears, you should not change gears while underway. The only exception is the highest road gear, when some forward motion from a lower gear is helpful. Like a car's transmission, first gear is a high-torque, low-speed gear, with higher gears decreasing torque and increasing speed. The top gear is often only a road gear (not used in the field) and speed in that gear is normally 9 to 15 miles per hour.

Note that most antique tractors do not have transmissions that are synchronized, so care should be taken while shifting. Occasionally gears may be very difficult to engage while the tractor is in motion. This leaves you in neutral, since taking the transmission out of gear is always easy; but then you cannot get it into a gear. If you happen to be heading down a hill at the time, you may be in for quite a ride since you will not be in gear and most antique tractors have weak braking systems.

You will find that first gear is handy for many applications requiring finesse, such as culti-vating, tilling, or for true lugging capability such as pulling heavy loads. I use it as a training gear for new operators and use it almost exclusively on the grounds at antique tractor shows and parades. Typically, however, you will find first gear one of your less-used gears. It is always a safe first choice, though, for any situation you are unsure of. Just realize that some implements, such as plows, will often perform better with a little more speed than first gear affords.

Second through fourth gears tend to be bread-and-butter gears that get your work accomplished. Mowing is done in second gear, or often third. Plowing is also often done in second or sometimes third gear, as the circumstances dictate. Higher gears are great for pulling wagons, manure spreaders, fertilizer distributors, and sprayers (rate of application usually determines ground speed). Fourth and fifth gears are less used but can be handy for finish work on very smooth fields, such as chain harrowing fields. The highest gear is often a road gear and is inappropriate for the field, or in the case of some tractors such as the Silver King, this gear is a true road gear and is downright dangerous to select in a field.

Additional gearing often exists, as do high and low shifters, and add-on transmission options, such as IH's creeper gears or the Sherman transmission add-ons for the Ford N Series. Gear selection is important to productivity, but err on the safe side and choose a lower gear if you are unsure of the correct gear to use. The higher gear will be there if you decide later that you need to use it.

Quick rear wheel width adjustments can be made with the adjustable rim, which is better known as the spin-out rim. By loosening locking bolts, you can rotate the wheel in or out to get the desired width. This is much easier than adjustable hubs on axles or dismounting and remounting wheel discs and rims to alter the wheel width.

Braking

Braking with antique tractors is quite different from anything other than heavy construction equipment. Because of the torque they generate, tractors can outpull their brakes. If you leave the tractor in gear, the tremendous torque of the drivetrain is simply too much for the brake design of 50 years ago. They will, of course, if in good repair, stop the tractor well if the clutch is disengaged. If the tractor is left in gear and the brakes are applied, they will offer significant resistance, and this fact makes them well suited for a second, but equally important purpose: brake steering. The brakes of antique tractors are independent, which means there is a separate pedal for each rear axle. Applying brakes individually can assist in steering by tightening your turning radius. Applying the right brake only, the tractor will turn to the right, and vice versa. This should only be done in the field under tillage at slow speeds: On the road this will create a dangerous swerve.

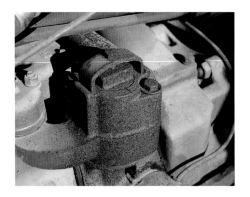

On many models of antique tractors, an air intake warmer, such as this one, was helpful in cold climates. Since all of them used exhaust air to warm the intake air, they all were prone to rust to the point of disrepair. This particular warmer on John Deere tractors was especially bad to rust, which rendered the gate (located under the small metal arch that looks like a knob) inoperable. Repairs can often be done, but regardless of repair or replacement, once it is working, oil it very frequently to prevent the assembly from becoming inoperable again.

CLUTCH

While the majority of antique tractors have the ubiquitous single-stage foot clutch we are all so familiar with, many do not. In fact, some of the most popular antique tractors, including the John Deere letter and early numbered series tractors, have hand clutches. Many other tractors had hand clutches as well, such as the Massey-Harris 55. While hand clutches operate on the same idea as foot clutches, their mechanical characteristics and feel are quite different. To operate a hand clutch, you must get used to two things: the ambiguous feel and the mental discipline and presence of mind to react with your hand in an emergency instead of your foot! Hand clutches operate differently from make to make, but they typically engage by pushing the lever away from you. This will have a smooth, progressive feel much like a foot clutch. Disengaging the clutch is accomplished by pulling the lever toward you, but will have a distinct "snap" to the movement if the clutch is not very badly worn or maladjusted. If you are unfamiliar with a hand clutch, I recommend several test drives and light chores in open areas before moving to more demanding operations.

The two-stage clutches are common in many 1950s-era and younger tractors that came with "live" PTO or hydraulics. On a live PTO, the tractor drive clutch can be disengaged, yet PTO-driven accessories will still be in operation. This is useful for chores such as running a baler, a post hole auger, pull behind combine, and so on. The two-stage clutch lets you disengage the transmission to stop forward movement about halfway through the clutch pedal travel, while the engine continues to drive the PTO or hydraulic pump. This becomes disconnected from the engine only when the clutch is depressed all the way.

There are many other clutches to be found on antique tractors and related equipment. Friction inertia clutches can be found on implements and stationary engines; wet multidisc clutches can be found as accessory clutches for hydraulic pumps and PTOs. Third-party add-on clutches, such as M&W's hand clutch for Farmall Ms and Hs, may be installed as well. The primary thing to remember is that all clutches are meant to be eased into engagement, and a maladjusted or badly worn clutch can do much damage.

Road Travel

Perhaps the most dangerous place to operate a tractor is on a public highway. While obvious dangers such as speeders and impatient motorists making foolish and dangerous passes come to mind, there are other problems. Many rural highways have limited sight distance. The speed difference between a car and tractor can make recovery from belated notice of the tractor nearly impossible for the motorist. Also, many towed implements are often wider than

Antique Tractor Tip

Dyno Your Dinosaur
At some antique tractor shows, a dynamometer is available for use by exhibitors. Try to take advantage of this machine, if possible, as it will give you a good idea of the types of horsepower your tractor is capable of producing.

the travel lane or pose a rear-end collision risk. The best way to avoid these dangers, of course, is to minimize the time you spend on public roadways. Another way to reduce risk is to pull off the road from time to time to let the accumulated traffic pass. This is common courtesy and also in the tractor operator's best interest, as impatient motorists are more likely to engage in dangerous behavior.

Putting an Antique Tractor to Work

Whether you wish to take advantage of it or not, an antique tractor is capable of performing serious work. Tractor work may be divided into four categories of field loads: tillage loads, towed loads, driven loads, and stationary loads. The first encompasses the majority of traditional implements, including plows, cultivators, subsoilers, and so on—implements designed to dig the earth and change the soil structure. The second category, towed loads, includes implements such as hay wagons, fertilizer distributors, and manure spreaders. Driven loads cover mowers, bush hogs, balers, post hole diggers, and the like. The broad category of stationary loads applies to work performed by a variety of implements, such as threshers, grinders, choppers, elevators, and blowers that are powered by the tractor's PTO shaft or belt pulley. The PTO or belt pulley may also be used to drive electric generators to power such devices as milking machines,

welding equipment, or household lights during blackouts. Powering water pumps and external hydraulic pumps for heavy-duty applications such as wood splitting is common as well. Less common stationary loads include powering rock crushing equipment or wood and shingle mills.

OPERATOR APPROACH

Operating heavy equipment safely and effectively is an art as well as a science and it doesn't play favorites. The father of a friend took some ribbing one fall because his wife was operating the combine while he drove the support trucks for the harvest. When asked why his wife was running the combine, he simply replied that she was better at it than he was. Becoming an ace operator has little to do with ego, mechanical ability, or a love for equipment; instead it requires juggling all the factors and circumstances facing the equipment and the task at hand in a fluid, safe, productive way that saves fuel and wear and tear on the equipment, and leaves the operator sane, in good humor, and with all appendages intact. In my friend's father's case, his wife could juggle ground speed, combine efficiency, "pass over" judgments (spots to miss for various reasons), fuel issues, and productivity better than he could. To become a good operator yourself, read your owner's manual, talk to and observe skillful operators, concentrate on the task and demands at hand, and practice. Antique tractors are not difficult

This is an unusual power steering unit available on the aftermarket for Farmall M tractors. It may be a Char-Lynn unit, but the owner wasn't sure. The steering wheel provides inputs (left or right) to the hydraulic unit, which in turn moves the control valve based on those inputs. Hydraulic pressure assists the operator in turning the front wheels.

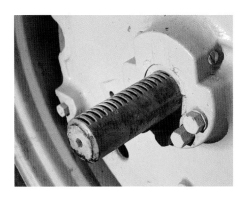

Adjusting the working width of the tractor's rear wheels is important with vegetable and row crop work. This John Deere's rear axle has a set of detents that helps the axle stay locked in place. There are many other designs. All of them require the owner to make sure the hubs stay tight and that both the hubs and the axles stay clean and in good repair to prevent the hubs from slipping along the axle.

to start, maintain, operate, or understand. But like most things, they do require a bit of study, some practice, a cautious approach, and the realization that good luck is no substitute for experience and training. Be safe, enjoy, and put that old iron to work.

CHAPTER 6

Safety

This antique tractor is set up with an overrunning clutch, which is an important safety accessory to any tractor without a live PTO. This device allows the inertia of the implement to overrun the speed of the tractor's driveline, thereby preventing the inertia from forcing the tractor forward. This is especially important when using implements with heavy, rotating components that spin at high speed, such as brush cutters.

While writing this book, I struggled with how to present safety more than with any other topic. Should I preach the extremes, about the pain and suffering of families who had lost their loved ones? Or should I presume you already understand how dangerous these machines can be and only point out less obvious safety issues? Should I take the middle road, using a mixture of both approaches? Should I present issues of safety throughout the book or confine them to one chapter?

My dilemma was solved when I realized that all approaches should be used. Throughout the book, I have provided safety tips to warn you of any danger relevant to the topic at hand. In a few places I discuss safety more vigorously, especially where I am covering risky procedures or discussing a subject I believe readers are likely to go to first. This chapter addresses safety alone, putting in one place all of the important considerations I know for minimizing risks when working on or around antique farm equipment.

Safety: An Approach and a Philosophy

Assuring your own safety requires more than making a few checks or memorizing a few rules. Acting in a safe manner requires thinking about safety every time you work on or near your tractor. The weight and power of a tractor, its high center of gravity, and the speed and sharpness of its implements can cause serious injury or death in the blink of an eye. Unseen or uncontrollable factors can cause risks for experienced and highly skilled operators, as well as for novices. Safety therefore cannot be an issue you think about from time to time throughout the year; it must be a philosophy, a way of approaching your tractor every time, with an eye and a mind toward eliminating risk.

As the operator of dangerous equipment, you must also be concerned about the safety of others. People who know and respect you may learn by watching you. If you act in an unsafe, careless, or reckless manner, friends or relatives might take your lead. Further, an accident the operator causes could injure other people. I only became safety conscious when I realized that I couldn't look myself in the mirror if I hurt others in an accident I could have done more to prevent. I am the one who is responsible for safety around my equipment and I do everything possible to provide a safe environment for the operation of my antique tractors. I hope you adopt the same outlook.

SAFETY BEGINS WITH AN INSPECTION

Safety begins with understanding the condition of your tractor. Try to get into the habit of making regular inspections for gas leaks, structural cracks, tire damage, loose bolts or fittings, and so on. These are not things that you will typically notice simply from operating the tractor. A friend in Elkin, North Carolina, nearly rolled his tractor when the downhill rear tire blew while he was bush-hogging his pasture. A simple inspection would have revealed the cut on the inside of the tire that caused the problem. A mounting plate that is loose and comes off at a critical time or a structural crack that finally gives way can cause the tractor to behave unpredictably, with any number of unfortunate consequences. Inspections are the only way to spot and redress potential problems before damage is done.

Safety Issues During Operation

HITCHING

A mishitched load is one of the most common dangers involving tractors. A drawbar load hitched to any other part of the tractor is a death trap waiting to snare you. This accounts for almost all rearward flips—in which the front end comes up and over and the tractor lands on the operator. Please note that drawbars such as those installed between the lower links of a free-floating three-point hitch are inadequate. Use or install

A set of fire extinguishers are a must for every shop. On the left is a carbon dioxide extinguisher that is rated for type B and C fires. These extinguishers don't leave a residue and are handy for electrical fires and small fires. The extinguisher on the right is rated for use on A, B, and C fires. There are five classes of fires, including class D and K, but these last two involve materials that typically don't exist in farm shops. When in doubt, grab the ABC extinguisher.

the tongue style drawbar that should be under the transmission case. Hitching the wrong implements or load to a proper drawbar is also very dangerous. Pulling certain loads can create what is termed "dead" loads, which stop the tractor abruptly, causing a rearward flip. An all too common example is a log being dragged through the woods. If the front of the log becomes jammed on a stump, rock, tree, or rut, the tractor can stop and flip in an instant. This very scenario kills dozens of operators every year in America. Unless you are absolutely positive you can assure your own safety, it is best never to use an antique tractor

to skid logs. Instead buy or hire a log skidder, a machine expressly designed for the job.

IMPLEMENTS

Many implements cause hair-raising safety concerns of their own. In particular is the danger of rotating PTO shafts. Spinning shafts often grab loose pant legs, sleeves, ties, and hands. Always keep and maintain guards on all spinning shafts and belts and stay far away while they are spinning. Never try to service or adjust these shafts and implements while they are in operation. Mowing and cutting implements, such as mower decks and brush hogs, should never be approached, nor used as a step, while they are moving! A man in our local area—an accomplished farmer—lost his foot when he stepped on a spinning mower deck. Due to rust underneath that was not visible from the top, the surface gave way when he stepped on it.

All implements carry a risk, but the risk can be minimized or removed through knowledge, keeping guards in place, and operating the implement only when the tractor operator is in the seat controlling it and the tractor. Remember as well to keep bystanders at a safe distance and only let someone approach when the implement has stopped moving.

STOPPING

Antique tractors often have notoriously poor brakes. Though slow speeds mitigate this problem some, failed brakes cause injury, damage, and death every year. Pay particular attention to downhill slopes while pulling a wheeled load, such as a hay wagon. Tractor brakes are often barely able to handle the tractor alone, and the addition of a wagon or trailer and a downhill slope can push them beyond capacity. Keeping the tractor in gear is a good way to supplement braking power.

A "flywheel" load driven by a non-live PTO or belt pulley can also overcome your tractor's brakes. Some PTO or belt pulley loads, such as bush hogs, have inertial forces that will continue to drive the tractor forward via the PTO shaft even when you depress the clutch and apply the brakes. You would be surprised at how many deaths and other accidents this scenario has caused. If you have

a non-live PTO, please install an overrunning clutch on your PTO shaft. Overrunning clutches are available at most farm and tractor supply businesses.

STEERING

Several conditions can reduce feel or precision in steering and create safety problems. Any significant load on the tractor, such as a heavy drawbar load, can cause light or unresponsive steering. If you are using a deep tillage implement such as a chisel plow, you may hardly be able to steer with the steering wheel at all, and you may have to rely heavily on the brakes for steering. A worn steering system can cause the front wheels to wobble at road speeds, which can lead to loss of control in extreme situations. Adding weights to the front of the tractor will restore steering feel and reduce wobble. A final steering concern involves narrow-front row crop tractors, which can sometimes roll over if a front tire blows at road speed. This is especially true for such tractors with only one front wheel.

FLIPS AND ROLLOVERS

Although a tractor can overturn far more quickly than a person would expect of an otherwise slow-moving machine, there are a few things an operator can do to keep all wheels on the ground. A sideways rollover can often be avoided by turning the steering wheel to the downhill side as soon as you feel the uphill side start to become light. Lowering a rear-mounted implement can also keep a tractor on its wheels. If your antique tractor has a hydraulic system with down pressure (many do, except Ferguson-style systems), dropping the implement quickly can stop a rearward flip. This accident is the number-one reason why top links on three-point hitches should be present when operating a three-point implement that doesn't particularly need one, such as a brush hog. Without the top link, the implement will fold over on top of the operator if the tractor flips.

Sometimes the safest course is simply to hold on and ride out trouble. For example, a downhill slide if the wheels are pointed downhill may make for nothing more than an adrenaline rush. If there are no obstructions at the bottom, no inconsistencies in the surface on the way down, and the distance is not long, then the odds are good that the tractor will make it to the bottom upright, with a frightened operator left otherwise unscathed in the seat.

HORNETS AND BEES

Nearly every year I hear of a death caused by a hornet's nest. The operator bails out, and is run over by his own equipment or is distracted and has an accident. In a recent case here in North Carolina, the operator was so busy swatting bees that he ran over a small rock wall that flipped the tractor over, killing him. Operator reaction would have made the difference between 50 bee stings and death. I have been told that the best reaction in this case is to set the parking brake,

shut down the implement if it is being powered by the tractor and then leave the tractor running. Apparently the heat, vibration, and noise from the tractor will keep the attention of some of the hornets, reducing the number that follow you.

How you react in an accident will often make a difference in the outcome. Think about the dangers listed here and other possible dangers, and rehearse in your mind how you will react. Expect the worst and have an escape plan.

Safety Devices

One advantage modern tractors have over antiques is that they were designed to incorporate safety features. Antique tractors are typically devoid of safety systems, and you must install them if the tractor is to be used for any field work. In fact, certain laws and regulations, such as those of the Occupational Safety and Health Administration (OSHA), require you to add certain safety features if you use an antique tractor for commercial purposes. Contact your

Jack stands are a great way to support your tractor if they are rated for the weight you'll ask them to bear. When removing large heavy assemblies or parts or trying to support the tractor on non-concrete surfaces, cribbing the tractor may be a better alternative. Be sure to check the engine removal photos for an example of cribbing. These two 3-ton jack stands have plenty of strength and safety for this situation.

state and federal labor inspections department for the current set of regulations and rules regarding safety systems on tractors.

Antique Tractor Tip

Tractor Roll Bars
A rollover protection system (ROPS for short) is required to minimize the chance of injury or death during a rollover. You should hire a local fabrication shop to design and install a ROPS for your tractor if you plan on using the tractor for any productive work.

One of the most important devices to help protect the operator during an accident is a roll-over protection system, or ROPS. This system consists of a structure that extends over the operator to protect him or her during a roll, and a restraining device to keeps the operator in the seat and within the confines of the roll bar. In modern tractors with cabs, all of this is designed into the cab structure. In other modern tractors the ROPS consists of the roll bar and a seat belt. A frame and front end strong enough to withstand rolling forces is also part of the system. In earlier years, before an ROPS was legally required to be present at delivery, an ROPS was an option. Tractors were still designed with these systems in mind, but the systems were sold separately as dealer-installed options or third-party products.

Many owners of antique tractors have been led to believe that an ROPS is a simple bolt-on accessory that they can add to their tractors or even make from scratch themselves. Unfortunately, this isn't the case. Simply bolting a roll bar onto an antique tractor may not protect you in a rollover! The axle, transmission, frame, and front ends of antique tractors were not designed for these systems and the forces generated in a rollover may cause the axle housings, attachment points, transmission housing, or frame to fail. If you wish to or must install an ROPS on your antique tractor, please be sure that the system was specifically engineered for your tractor. If you cannot buy one designed specifically for your tractor, ask any full-service metal fabrication shop to build one for you. They will design one, or hire a mechanical engineer to design one, that they feel confident will protect the operator in all circumstances.

A kill switch, which stops the tractor's engine in certain circumstances, is another common safety device and comes in several designs. Seat-based kill switches and mercury-leveling kill switches may be used for antique gasoline engine tractors and for antique diesel tractors retrofitted with electrical fuel supply cutoff. Seat-based kill switches cut the engine if the operator leaves the seat of the tractor. Common on modern tractors and lawn equipment, a seat-based kill switch is a sensible modification to an antique tractor. This retrofit should be designed so it can be bypassed by mechanics and others who have a legitimate need to start and run the tractor without being in the seat. A mercury-leveling kill switch will kill the tractor's engine if the tractor starts to flip. It completes an electrical connection when the tractor reaches a certain angle. In hilly areas, these devices can be a nuisance at best or useless and dangerous at worst. Additionally, they can be dangerous in loading the tractor on a trailer. While mercury switches have some benefits, they are not appropriate for all conditions.

Antique diesel tractors with a mechanical fuel supply cut-off can be retrofitted with a marine style "man overboard" key. With this system, you reverse spring load the cutoff lever and place a "key," actually a pin, in the lever. You then connect a lanyard to the pin and to a strong clip that you attach to your clothing. If for some reason you are thrown from the tractor, the pin is pulled and the shut-off lever cuts the fuel supply. These systems are not commercially available that I know of and must be custom made. Any mechanic or marina should be able to help you if you aren't comfortable designing this yourself.

Other safety devices that should be or can be present on an antique tractor include a fire extinguisher and a spark-arresting exhaust system. One of the design flaws in most antique tractors is the gas delivery system. These are notorious for leaking, and the gas tank and the leaky sediment bowls are almost always located over major electrical components such as the battery, dash, voltage regulator, and so on. The extinguisher should be readily accessible to an operator or a mechanic standing on the ground working on the tractor. A spark-arresting muffler will reduce the risk of fire in dry conditions, in which a single spark can light up a barn or field in a heartbeat.

Safety in the Shop

Safety in the shop—the place where you maintain and repair your tractor—is as important as safety in the field. In fact, shop accidents account for more injuries each year than accidents involving the use of tractors. General rules for shop safety apply to any type of workshop. I will not go into the basic rules, like wearing eye protection when drilling, hammering, or sawing, and wearing a mask when painting. If you need basic guidance on general workshop safety, ask at your local library, or local hardware or building supplies store, what safety references they can recommend.

Keeping your tractor workshop safe begins with realizing three things:

1. Most parts of antique tractors are very heavy.

2. Most fluids used in the operation or repair of an antique tractor are explosive, flammable, poisonous, carcinogenic, acidic, or highly caustic.

3. Power tools, hoists, and other equipment needed can be dangerous to use.

Antique tractor parts can be quite heavy. Lifting such parts improperly, trying to lift too much weight, or catching your fingers underneath or between

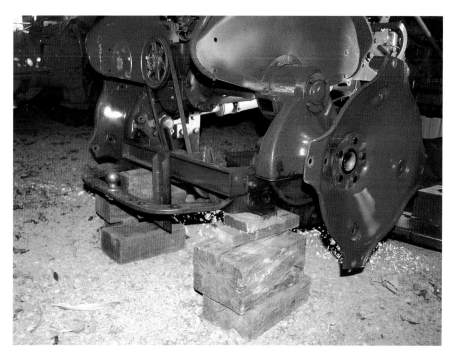

Here is an example of how to safely crib a tractor. The blocks are large and heavy, and the corners of the hitch frame, which are always securely attached to a tractor, are the crib points. Cribbing is especially useful on gravel such as this where jacks stands don't have a proper bearing surface.

heavy components can lead to injury. Use a sturdy cart, dolly, hand truck, or hoist for lifting or supporting components that could be heavy. The rear wheels, for example, are usually much heavier than they appear. Most of us lack the strength to stop a rear wheel loaded with ballasting solution and a wheel weight from falling once we remove the last bolt or slide it from the axle, so attach a hoist first. Be careful, too, how you store tractor parts. Storage shelves must be especially strong, as the accumulated weight of all the parts will reach hundreds of pounds.

Another dangerous operation in the shop is that of "splitting"

Antique Tractor Tip

Climb On with Care
Believe it or not, the simple act of climbing on and off a tractor, especially when it is wet, is a common source of injuries. Always use caution when climbing on or off equipment.

This is a WorkSaver brand aftermarket three-point hitch system. Note the hand wheel welding to the center link. This makes adjusting the center link much easier. It can even be adjusted from the operator's seat. Whenever you reach down toward an implement to make an adjustment like this, make sure the tractor's PTO is off and any moving parts on the implement have come to a complete stop.

the tractor to perform engine or drivetrain work. Many antique tractors have no true frame, relying instead on the strength of the transmission casting and the engine block for support. When these tractors require

clutch repair or significant engine or transmission work, the main engine and transmission assemblies must be unbolted from each other, leaving the tractor in halves and therefore unstable. You may also need to transport

one half, or to lift and support it to perform repairs. The support and transportation system you use depends on the operation you are performing, but there are a few ground rules about supporting and transporting parts of a tractor that are universal:

Tractors should only be supported by overhead hoists, jack stands, or "cribs." Cribs are large-dimension wood sections (6x8x24 inches, for example) laid together log cabin style. They make an acceptable substitute for jack stands, and in some situations such as field repairs where jack stands may be unstable or sink in the soft soil, they are preferable.

Each jack stand must be rated to support at least 3 tons for the average tractor. While the load on a jack stand may not be that much, jack stands rated below this tend to be hobbyist or automotive grade stands that can be dangerous, especially if the load tried to shift or move slightly. Some mechanics will not use any jack stand rated below 6 or 8 tons.

EVERY support operation must have a backup. Beside each jack stand and underneath every hoisting operation should be a crib of wood to catch the tractor, should the support fail.

Moving large assemblies or halves should be carefully orchestrated and engineered. The assembly must be supported by a movable system, and this adds additional risks. If the assembly is to sit on a dolly while being moved, then you should be sure there are at least four casters (three-caster systems are too dangerous) and be sure the assembly is bolted or strapped to the dolly so it cannot shift during movement. If the assembly is to be moved while being supported by a hoist or crane, try to pull the hoist or crane by using a leash or rope, so if anything fails, you are clear of the assembly. Consider also a support underneath to minimize risk or damage to the assembly.

Permanent supports for the tractor and its assemblies should be of good quality and their use should be planned ahead of time and carefully executed. Engine stands should not be overloaded and transmission jacks and other permanent repair stations should be overengineered so they can serve you well in all circumstances. Permanent hoisting is typically frowned upon unless you have designed your building specifically for these dangerous dynamic loads. Most older shops and buildings are not structurally capable of supporting overhead hoists. Also, hoists tend to be limited in their usefulness since they only run in one direction, requiring you to line up your equipment and work in a certain area. Movable cranes and hoists are adaptable, less expensive, and can be taken apart for storage. Remember, in a pinch, the front loaders on tractors and backhoe buckets make handy dependable, movable lifting systems. Just remember to keep clear of the load and use these temporarily and not for long-term support.

Most all of the supplies used in a tractor or on a tractor are dangerous. Antifreeze is toxic, gasoline and diesel fuel burns

Antique Tractor Tip

Extinguish the Risk
Fire extinguishers are easily added to antique tractors and you should seriously consider adding one. Choose a mounting location that is easy to reach from both the operator's seat and from the ground.

with a vengeance, and paint is carcinogenic or poisonous or both. The old paint on the tractor is lead based. Lacquer thinner shows up in your liver within five seconds after it has splashed onto your skin. Central nervous system damage comes from overexposure to any number of chemicals in the shop. Used oil has a high heavy metal content and lye-based cleaners cause burns and blindness. Piles of oil-soaked clothes and rags can spontaneously combust, causing a fire. Obviously, gloves and safety glasses should be worn around any of these fluids, and respirators and dust masks must be used as needed. Fortunately, all of these chemicals are sold with material safety data sheets, which are not automatically given to you in most cases, but must be available by law. Ask for them when you buy the products. These sheets contain important safety and emergency information you need to use these products safely.

We all know that power tools can be dangerous to use and cause many injuries. Most arise from disobeying basic safety rules, like

wearing goggles, keeping hands and fingers away from spinning equipment and blades, and so on. One of the most important safety rules is always to use power equipment in a comfortable posture and at a height that allows you complete control of the tool. Remember that each power tool comes with an operator's manual that outlines the safety precautions that must be taken. In fact, some manufacturers include these on a placard that can be hung on a wall as a reminder.

These basic safety precautions must be observed in every situation if you wish to have some assurance of your personal safety. If you are uncomfortable about a particular application of a power tool, then you can rest assured it is unsafe. The reverse is not necessarily true, though: Being comfortable is not a prediction of safety. Complacency is one of the reasons safety rules are broken. Always second-guess yourself and make a conscious attempt to remain aware of safety issues while operating a power tool. Provide adequate lighting to the work area, and keep the equipment in good repair.

Hitches and Implements

This is a typical hay tedder, an implement to turn and re-throw hay to improve the drying speed and consistency.

Having a tractor at your disposal and in good running condition is often only half the battle when it comes to actually putting your old iron to work. Antique tractors, particularly those built before the 1950s, have a number of systems that are not standardized. The next trick is finding implements and attaching them to your tractor, which may require some modification. Read on, and we'll sort through some of the things you may need to do to get out plowin', mowin', or haulin'.

Likely Retrofits You May Have to Make to Your Tractor

While antique implements designed for your tractor are available, odds are you will run across a use for your tractor that no antique implement, for whatever reason, is available to fill. At that point you will be faced with using a modern implement. While modern implements are often simple and require no special adaptations to your tractor, most will require a three-point hitch. These hitches can be custom-made locally or bought as a kit you can install on your tractor. The kits are available only for the most common tractors, however.

Other retrofits to consider include an ROPS for safety, auxiliary lighting, and a 12-volt, negative ground electrical system. ROPS, discussed in the previous chapter, is sometimes available as a kit specific to your tractor; otherwise a metal shop can design and build one for you, and assist you with installation. Auxiliary lighting is often necessary, as many antique tractors didn't have lighting at all, especially rearward pointing PTO lights that illuminate an implement or machine powered by the PTO. Twelve-volt electrical systems are sometimes necessary for brighter lighting or modern electrically powered auxiliary systems, like radios and global positioning systems (GPS).

Whatever the reason, modifying farm tractors to fit the use is a time-honored tradition in rural America. Do realize, however, that permanent modifications can detract from an antique tractor's collector value. Consider bolt-on retrofits that can be removed if you decide you want to sell the tractor for its value as an antique. A good metal fabrication shop should be able to help you design and build any retrofit you need.

HITCHING SYSTEMS

Throughout the history of tractor manufacturing, designers have worked to develop and enhance safe, secure, and versatile hitching systems. Tractor hitching systems fall into three broad categories: drawbar, proprietary, and industry standard.

Before three-point hitch systems were invented, each tractor manufacturer had its own hitching scheme. Farmall had several and Allis-Chalmers had the Snap-Coupler. Here is a plow designed for use with the Snap-Coupler hitch. Implements for all manner of proprietary hitching systems are becoming harder to find.

Trailer plows such as these do a much cleaner, consistent job of plowing fields than three-point hitch-mounted plows can manage. These classic John Deere plows are among the best ever made and if you need a plow, I suggest you buy a trailer plow like these rather than a three-point hitch plow.

Average size antique tractors with a three-point hitch, such as the Ford 8N, MF40, John Deere 530, and others, can very successfully drive 4- to 5-foot roto-tillers. This makes very short work of gardens and new yard space. They are priced about 20 to 30 percent more than walk-behind tillers and till 100 to 150 percent more ground per pass.

Finish mowers are another great modern implement that antique tractors superbly handle. This 5-foot finish mower could be driven by any mid-sized antique tractor with a three-point hitch.

Here are two methods of fertilizer distribution. The ground-driven (left) method more accurately distributes the fertilizer while the broadcast (right) is quicker. Broadcast distributors can also be used with seed while ground-driven distributors can not. I find the ground-driven distributor much easier to load and neater and cleaner to use (you don't inhale fertilizer dust), but I understand the attraction to the broadcast spreader. In the end it's your choice, and rest assured an antique tractor can handle either without any problems.

This three-point cultivator is sized for a large garden or small vegetable farm. This three-point cultivator and single bottom 12-inch plow are sized for a large garden or small vegetable farm. These would work well behind a small antique utility tractor with a three-point hitch.

Antique Tractor Tip

Implement Insight
Unless you have a good supply of antique implements and plenty of spare parts, three-point implements are typically the best choice for regular production work.

Drawbar implements were the first implements designed for tractors and this design is still used for wagons, manure spreaders, light-duty chain harrows, and so on. Often, horse-drawn implements were modified to be pulled by tractors at the drawbar. Drawbar design on antique tractors varies widely, from a mere tongue extending from under the rear axle, to an elaborate floating bar, such as was available on John Deere MTs.

Proprietary hitching systems were peculiar to each manufacturer. Some examples of proprietary systems are IH's Fast Hitch and Allis-Chalmers' Snap-Coupler. These systems were robust and were initially popular with customers. The drawback to these systems, however, was that they limited the tractor owner to implements designed for his unique style of hitch. Typically this trapped the farmer into using implements by his tractor manufacturer. He couldn't share implements with his neighbor or even with another tractor of his own if the hitching systems differed. In addition, certain implements were made only for certain models of tractors (for example, a plow for a Farmall A will not fit a Farmall C). Under the proprietary system, the farmer was, in essence, betting the farm on a particular style of hitch. This clearly wasn't acceptable and created the need for an industry standard hitching system.

The hitching system that became the industry standard is the three-point hitch, invented by Harry Ferguson. His design first found a home in North America on the Ford N Series tractor. This system had several advantages over other systems, as it was hydraulically controlled and would also react to field conditions to alter the presentation of the implement. While some manufacturers had hydraulic systems that could raise and lower an implement, they did not react to field conditions and required the operator to make constant adjustments. Although Ferguson's system had its detractors, and the design did have some weak points (no down pressure, for example), the system was a good mix of a lot of attributes needed for the next generation of farming. When Ferguson's patent expired in the 1950s, most manufacturers adopted the three-point hitch. Among the first to do so was Massey-Harris, with which Ferguson merged in 1954. IH and Allis-Chalmers waited until the 1960s to offer it. Ferguson's hitch was immortalized when the American Society of Agricultural and Biological Engineers (ASABE) adopted it as the Category I three-point hitch standard.

HITCHES, IMPLEMENTS, AUXILIARY SYSTEMS, AND FIELD METHODS
The Hitch Classes and Their Importance

Antique tractors typically use a hitch from Category I, II, or III, as set by the ASABE. These are not physical hitches that you buy, but are hitch specifications, or standards. There are many other agricultural hitch specifications. For example, Category IV hitches are designed for tractors with more

Many antique tractors have three-point hitches for a round bale spear such as this, but be careful. Round bales are heavy and many lighter utility tractors with three-point hitches, such as Ford 8N, typically don't have the hydraulic strength to handle it. Many tractors also don't have the weight or stability to carry the bale. Leave this job to the larger antique tractors or your modern tractors.

This is considered by many to be the gold standard for hay rakes. It is a New Holland ground-driven hay rake and most any antique tractor with more than a 12 horsepower rating at the drawbar can operate the rake.

A boom pore attaches to the three-point hitch of a tractor and helps lift heavy objects. Be sure your tractor can handle the weight of the object before attempting to lift and carry it.

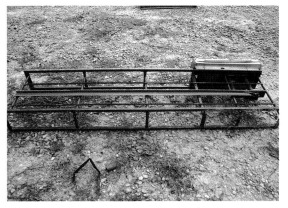

This is the main section of a hay elevator and contains the drive gear and the pulleys. Additional sections are added until you have the length you need, and then a special hay conveyor chain and a motor are added to the main section of the elevator.

Antique Tractor Tip

Implement Scrap Yards
If you need a part for your implement, used implement yards are available, though they are not as common or organized as used tractor part yards. Some commonly worn components like chains, plowshares, and so on, are widely available through many mail-order companies and local suppliers.

Post hole digging requires a three-point hitch with down pressure and live PTO, which is something many antique tractors don't have. Some tractors, such as the Massey-Harris two digit series (22 and 33) and many John Deere tractors, do have live PTO and down pressure and would operate this post hole digger with ease. This implement can be dangerous with any tractor, antique or modern, so use it with care.

Many implements have a custom mounting system and mount under the belly of the tractor. Mowing decks commonly attach this way. This is a Woods mowing deck under a Farmall B.

A brush guard is a great addition to any antique tractor being used on a regular basis, especially tractors used to operate near wooded areas, such as a tractor used for brush cutting. These will protect the radiator and grille from trees, fence posts, and larger branches.

horsepower than antique models, and Category 0 is a very small specification commonly retrofitted to small models, such as the Farmall Cub and Massey-Harris Pony. Adapter kits to use Category I implements with Category 0 hitches and vice versa are available.

The most common hitch specification found on antique tractors, by far, is the Category I hitch specification. This hitch standard is based on Harry Ferguson's design, and any 1939 or newer Ford tractor, and any Ferguson tractor, will have this hitch. Category II hitches were based on John Deere's hitch design, which was modeled on Ferguson's. This hitching system is very strong and is common even on today's tractors. It's rated for tractors with 35 to 75 horsepower. There is a "narrow" version of the Category II hitch design that is handy on row crops and tractors with narrow rear-wheel spacing. This is not common, however.

Regardless of the hitch type, it is up to the operator to determine if the tractor can actually safely use a particular implement. Many implements that fit your hitch may be too large or otherwise unsuitable for your tractor and cannot be used safely!

What Type of Implements: Modern or Antique?

Proprietary implements that were available with your antique tractor when it was new can be restored and used with your tractor. The problem is these implements can be scarce and expensive. In addition, parts can be difficult to find, especially renewable parts such as shares, disc bearings, and the like. Even so, many with an antique tractor hobby enjoy collecting, restoring, and using these implements, which usually perform beautifully. In fact, some antique proprietary implements outperform their more modern three-point counterparts when completely refurbished and properly mounted. Antique implements should not and cannot be used in every situation, however, and the trouble and expense of acquiring and restoring them can be more than some can justify. This is especially true for folks who are using their antique tractors for modern production.

If you plan on using antique implements, then your next steps seem obvious. Just find the implement, fix it up if necessary, and use it, right? Unfortunately, there are a million pitfalls along the way. First, finding one locally often seems to be impossible. I almost never see proprietary implements for sale locally when I need them. I almost always have to scour the national antique tractor publications and the regional tractor and implement junkyards to find what I need. That means shipping costs or a long drive to bring it home. The next pitfall is finding one that is complete. Many of the implements, especially those like mowers that have lots of parts, always seem to be missing some critical but easily overlooked part such as a bracket. In fact, if you are in a hurry because you need to press it into service on your farm, I would only buy the implement promised to be 100 percent

This is a typical hay wagon and is one most antique tractors would be adept at handling if loaded properly for the size of tractor being used. Antique tractors also often get called to perform hay rides, and while it is fun, safety is paramount. Hay wagons like this without sides are dangerous to use for hay rides. Use a sided wagon or fabricate sides for any hay trailer being recruited for hay rides.

Land levelers such as this are designed to help create fields with a uniform surface and slope. This helps prevent high spots, which are often dead zones for some row crops, and to help shed water evenly to prevent the formation of erosion gullies. The wheels behind the leveler are adjustable and dictate the working depth of the implement. Any of the larger antique tractors can handle this implement without a problem.

When shopping for a farm wagon or trailer, be aware that these are often available for sale as a bare frame and you can then build exactly the type of wagon you need. This four-wheel trailer with hydraulic ram for dumping and a steerable front axle would be great for moving bulk material.

Soil contact is critical for high germination and survival seedling rates for any small, shallow planted seed like hay, clover, or lawn grass seed. After broadcast seeding or drilling the seed, make a pass with this cultipacker to improve soil contact.

Manure spreaders are widely available as new and have changed very little in design over the years. This older New Idea spreader has some age on it but it's in great shape. Modern ones are built the same way and operate no better. There is no reason not to buy this used spreader and hook it to your antique tractor.

Implement yards, such as Leinbach's of Winston-Salem, North Carolina, usually have a wide variety of replacement and used parts. When your implements need a new blade, tooth, rod, disc, or bearing, yards like these are the place to get replacements. Here are stacks and stacks of replacement discs and coulters for disc harrows and disc plows.

complete by the seller. Otherwise you may spend weeks locating a replacement for a missing part.

Once you bring the implement home, it often can represent a restoration challenge in and of itself. Again, implements such as mowers, manure spreaders, and the like that have many renewable and rusted parts can be very expensive to refurbish. A new set of guards, ledger plates, knives, and hold-downs for a small 5-foot sickle bar mower can cost well over $100. Then mounting up the new implement can be a real challenge. Mounting a plow for a Farmall model A, for instance, is not an intuitive process and can present a real challenge to someone new to the task. Some implements may require two or more people to

mount. While I do not mean to discourage you from using antique implements, be aware that they represent their own set of challenges and can be more costly than their modern cousins, once you have figured all your costs to acquire and restore them.

Using a modern implement can be much simpler. Simply purchase one of the many commercially available three-point hitch retrofit kits made for your specific tractor by any one of several different companies, install it, and then use the properly sized modern implement of your choice. It is all very straightforward, unless there is no commercially available three-point hitch adaptation kit for your model of tractor. Even then,

however, most metal fabrication shops can design and build one for you. The main stumbling block for three-point retrofit kits is that these kits, whether they are commercially available or have to be custom-made, can be expensive and usually require hydraulic lifts. Most folks with some metal-working and welding skills can make their own to save money; but even then plan on spending at least $100 on metal, top links, and other necessary parts. The nice thing about these kits is that they are usually reasonably simple to mount to your tractor and are bolt-on accessories—that is, they do not require you to modify your tractor. That way you can remove the hitch if need be.

Putting Your Tractor to Work

Now that you have an antique tractor and are ready to put it to use, where do you start? More than likely there are specific chores you know you want to perform with the tractor and need a few ideas and pointers. If that is the case, feel free to skip to the section of this chapter that addresses that particular task. I do, however, want to encourage you to read this whole chapter as there are many ways to perform most field chores, and understanding as many field methods and implements as possible will help you to become more resourceful and effective. For instance, there are many implements to mow a field with, more than one way to cultivate row crops, and how you plant depends on a wide variety of factors such as the crop being planted, climate, topography, and so on.

Raising Crops

PREPARING THE SEEDBED

Whether you are planting hay, planting a row crop, or simply planting a garden, your first task will be to create a finely worked seedbed. Your options are endless and in fact,

The Farmall 560 is a workhorse that can be put to immediate use on any modern farm and could work right beside any modern tractor. Plenty of power and all the modern conveniences and power options, such as live PTO, torque amplifier, power steering, three-point hitches, power steering, cab, and more are available for this tractor. These were made in the late 1950s and early 1960s.

Antique Tractor Tip

Seedbed Prep

When preparing a seedbed, soil moisture is the most critical factor. If the soil is too dry it may be hard to work, and will never break apart as completely as it could. If it is too wet, traction is difficult and the soil will again be left in a poor state.

seedbed preparation has undergone a revolution of sorts in the last 10 years. If you are planting for market or for livestock production, you may want to talk to your Agricultural Extension Agent about the recent developments in no-till planting. If you are not planting for market, are planting just a small area such as a garden, or simply prefer to avoid the use of herbicides that no-till requires, then your options boil down to just three: plowing, disking, or tilling. Of course, a combination of these approaches can be used, and in some circumstances they must be used together. Plowing primarily creates a deeply turned and worked field that typically requires additional work to make ready for seed. Disking also can be used to deeply work a field, but can also be used to create a field ready for planting. Tilling is a one-step process that creates a finely worked seedbed ready for planting.

Getting Ready

In preparation for plowing, the field should be grubbed for roots, stumps, and rocks. (If the land is freshly cleared, the plowing should be done slowly to avoid danger of equipment damage.) Also, a light disking before plowing can be advantageous in all conditions if feasible, especially if there are a lot of weeds and crop trash on the field. The disc will cut the surface trash and turn it in with the top soil, allowing easier, cleaner plowing. Plowing is often best done in the fall, so crop remains and weed remains have a chance to soften and break down over the winter. Turning over the remains of a crop and exposing the roots to air also helps to minimize certain types of pests that over-winter in the root structure. Depending on your climate, fall plowing allows inclement winter weather such as deep freezing and winter precipitation to create a very friable, easy-to-work soil structure that will disc well in the spring in preparation for planting. One negative to fall plowing in some areas of the country is soil erosion.

Plowing

An entire book can be written on the art of plowing, and I firmly believe that plowing is a skill that one can spend a lifetime perfecting. In addition, there are many different types of plows, though typically the "middle busting" and "turning" plows are the only ones used regularly. The turning plow does what its name indicates: It turns the soil over upon itself, leaving the slice of earth the plow cut upside-down. The middle busting plow does the same, but it takes two slices of the earth and turns them both at the same time, one to the left and one to the right. This implement gets its name because it is used often to lay open the middle furrow of the field. There are other plows, but these are the two most common plows and plowing methods.

To begin plowing, you must learn your options, make your choices, and then lay out your field with this strategy in mind. Plowing creates a heavily worked field that will literally turn the top 5 to 10 inches of dirt upside down. This will change the structure of the soil, help to kill weeds, and create a very rough field. It aerates the soil, and when used in conjunction with disking and other finishing tillage methods, it creates a very friable fine texture excellent for planting. The options for most plowing in North America are going to be using a turning plow or listing plow in typical flat-land plow patterns, or using a turning plow while plowing on contour.

There are other plows and other methods, but these are best researched in your own local areas with agricultural agents better suited to explain and show them to you. To choose between these two options, you should look to your land. Begin by imagining plowing a furrow parallel to the fall line of the land. If this imaginary furrow would become a raging river of mud in a heavy rainstorm, eroding your land in significant ways, then contour plowing may be best. At the very least, you should consider contour plowing the field in lands that at least angle significantly away from the fall line. Otherwise, typical flatland plowing would work for you.

Let's touch on how a typical turning plow works. A typical plow operates by slicing the soil and turning it over to one side as it leaves the slice upside down. Normal turning plows leave these slices turned over to the right (looking at the tractor and plow from behind), meaning that the unplowed portion of the field is always to your left. By logical extension, you can imagine that the last furrow cut will be left as an empty trench. This empty trench is called a dead furrow, and while dead furrows are unavoidable, you want to locate them strategically.

While plowing a furrow is a simple operation, laying out your field to begin with can be quite challenging. Will you plow on the contour, or will you lay open the field in large sections called lands? How will you actually traverse the field so your time and work is minimized? How will you lay out your headlands and will you plow them? (Headlands are areas outside of the plowed area where you turn your equipment around.) Into how many lands, or work areas,

Larger classic or late model antique tractors can still be called to work. This International is a fuel-saving diesel with a three-point hitch, modern operational features, and hydraulic ports. It has everything you need for a powerful workhorse on any farm.

will you break up your large fields? Exactly how you plow the field will usually not be obvious and a strategy must be developed. The method you ultimately settle upon is less important than realizing that you can't approach it haphazardly. Develop a strategy that meets all of your needs and stick with it.

Typical Plowing

Typical flat-land farming does not require paying attention to contours, and these fields can be laid out in ways that are simply the most efficient to plow. Usually the fields are plowed in a circular fashion, starting far enough from the edges of the field to allow an area in which to turn the equipment around. The first trip around the field is made clockwise at the inside edge of the headlands. (Refer to the diagram for more information.)

The clockwise direction is usually chosen because throwing the dirt in toward the field is slightly more advantageous than throwing dirt out away from the field. By throwing dirt in, you create a dead furrow around the field, giving you a reference point from which to drop the plow at the beginning of each furrow. This dead furrow helps to make the finish plowing of the headlands

Antique Tractor Tip

Low-Tech Tools
Sturdy rods, poles, and mallets are especially useful for mounting and removing implements, even three-point hitch implements.

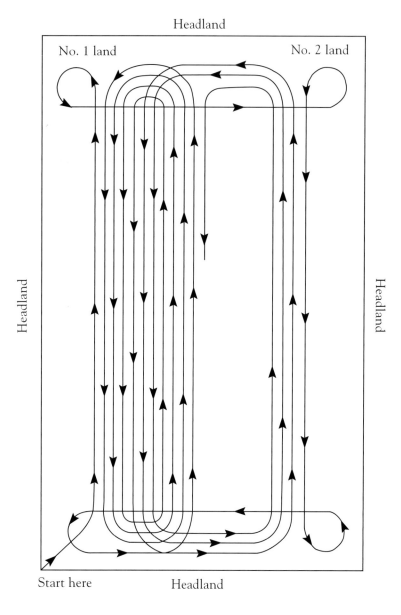

Headland

No. 1 land

No. 2 land

Headland

Headland

Start here

Headland

Antique Tractor Tip

Plow Adjustment Made Simple
After mounting a plow to your tractor and opening an initial furrow, find a flat spot of ground and lay down a block of wood whose height matches the furrow depth you have chosen. Then drive the tractor's left rear wheel up onto the block. Adjust the plow so the bottom frame of the plow is flat on the ground. Some fine tuning will be necessary once you start, but this will be close to correct.

easier as well. After that one pass is made, a trip is made down each side of the field in turn, starting at the outside edge and then plowing toward (but throwing the dirt away from) the middle.

As you continue, you approach the middle of the field from both sides, until there is too little distance between the furrows to turn the equipment in. At this point you would begin with the second land, plowing the far edge but returning up the middle of the first land. After a few passes, you have completed the first land and you now begin plowing the second land exclusively until you once again have too little room to turn and make the next furrow. You then repeat the process by starting the third land while finishing the second on the return trips. This continues until the field is finished.

The last step is to plow the headlands. To do this, you can either plow clockwise to continue the plowing pattern begun by your first trip around the field, or you can plow the other way. Continuing clockwise is the easiest and will more than likely be the neatest job, but because the first furrow made around the inside of the headland is now a jumbled mess from beginning all the land furrows at this first headland furrow, it is not terribly critical which way you do it. The only suggestion I have is to rotate direction from year to year. This will help keep the soil from migrating away from the field edges. If you always travel clockwise, you remove dirt from the edges of the field. This is also true of the field furrows. You will want to change directions from year to year to minimize or eliminate soil migration caused by plowing.

Contour Plowing

Laying out the field for plowing depends on several factors, such as the shape of your land, its drainage patterns, areas to leave fallow, and so on. Plowing on the contour is logically the easiest—but actually the most difficult—to do accurately. To plow on the contour you must create furrows that are perfectly level and make no change in elevation. Typically, this necessitates some type of layout stakes placed throughout the field to maintain consistency in elevation. With this type of plowing, throwing dirt uphill or downhill is the only decision to make. Most folks, if they have a need to plow on the contour, usually feel that they should throw the dirt uphill to counteract the previous year's soil creep and erosion. That means you would start at the top of the hill, and with a typical one-way plow, start on the right-hand side (looking at the field from the lowest point) of the field and plow to the left.

Anytime you must throw dirt in a specific direction through the entire field, such as contour plowing, or when you are plowing a small garden that doesn't lend itself to a rotating scheme, plowing time will be cut in half by using a two-way plow. A two-way plow is a plow that has both right and left bottoms. That is, the plow has bottoms that will throw the dirt in either direction, but only

Antique Tractor Tip

Three-Point Perspectives
If you plan on owning and using many different implements, consider buying modern implements and retrofitting your tractor with a three-point hitch. These implements are much easier and less expensive to find than original implements.

half the bottoms are used at any given time. Having both types of bottoms allows the operator to move back and forth across the lands from both directions, yet the dirt will always be thrown in the same direction. For example, the operator can plow heading north with the standard right-hand bottoms. This throws the dirt to the east. Then, instead of having to circle the field, he can pick up the right-hand bottoms, lower the left-hand bottom, and plow heading south. The dirt will still be thrown to the east. While on the surface this sounds very handy and ingenious, most fields are plowed with a rotating scheme that minimizes the problem associated with having a plow throw dirt in only one direction.

Some Other Plowing Considerations

One other important plowing operation for the arid parts of the world is listing. Listing is not technically a plowing operation, but a plowing-planting operation. Listing consists of creating deep, well-rounded furrows in a field

and then planting the crop in the furrows themselves. The deep furrows create a trap for what little rainfall does occur and the water is directed to the plants. Then during the year, the act of cultivation knocks soil from the furrow ridges and covers the plant roots even deeper, making the plants even more drought tolerant. Listing is less common in these days of irrigation and minimal till soil-management practice, but was an important method when antique tractors were new.

The lister is a device that looks like a two-way plow with the bottoms bolted together back to back. Occasionally, a planter follows a listing plow that plants

Antique Tractor Tip

Use Tractor Pulls to Fine-Tune
Antique tractor pulls are great places to experiment with wheel weights and tire ballast for maximum traction and stability.

the seed and slightly covers it. The farmer may list fields in the fall to allow snow and winter rains to moisten the deep furrows, and then in the spring he or she will list again, splitting the ridges and covering the furrows. This traps the winter moisture under the newly created ridges, creating a bit of an "aquifer" for the newly germinating seeds. The lister without the planting attachment is also called a middle buster or a potato plow. Technically these implements differ slightly, but the names seem to be used interchangeably across the country.

Plows are also handy for cutting trenches, making soil-erosion barriers, and deeply working additions such as lime, manure, and cover crops into the field. To cut a trench, simply make a pass up one side of the proposed trench's centerline and back down the other. This turns the dirt away from the centerline. While these trenches are seldom pretty, they will work well. Middle busting plows, or a plow called a potato plow, are especially handy for this. Creating soil-erosion barriers, especially water breaks, is a common use for plows. Here you plow two or three times on the

contour, following nearly the same centerline each time, throwing the dirt downhill. This creates a level transit through the field. Placed strategically on hillsides, these barriers prevent water from gaining any of the speed it needs to erode the field heavily. Turning over weeds to kill them, turning in a winter green manure or a nurse crop such as clover in the spring, and turning in slow-release products such as lime are other tasks for your plow.

Disking

Disking is another field-preparation procedure. Disking implements work by forcing disks (actually, "saucers" is a more descriptive word), lined along an axle that is canted to the line of travel, through the soil. The most common disking implement, and the implement folks are referring to when they say a "disk," is the disk harrow. The disk harrow has two or more sets of disks, called gangs, arranged to work in opposition to each other. There are other disk implements, such as the disk tiller, or stubble disk, disk plow (used as a plow), and finishing disk. In addition, there are many variations of these disks. The function they all perform is to take the soil, cut into it, work the slice, and leave a well-worked soil behind. A good disking implement also has the ability to cut and mulch surface trash and additives and work them into the soil.

These implements not only cut and turn the soil, but also leave a fairly smooth field, especially if you drag a smoothing harrow (or even a log) behind the disk. Disking is done after plowing or it is used by itself. If used by itself, the disking is

usually done in two passes. During the first pass, emphasis is placed on the cutting and turning action. The second pass is used to smooth the soil, leaving a suitable seedbed for planting. Sometimes different disks are used for each pass. When the soil does not need to be deeply worked, a disk harrow will create a seedbed faster than a plow and will cut field trash (weeds, crop remains) better than a plow. While the disk harrow does not replace the plow, it is now used in many of the circumstances the plow would have been chosen for in the past, and it has caused the use of the standard turning plow to fall dramatically over the last 30 years. The disk harrow is used almost exclusively on some farms and many now consider the plow to be more of a special-use implement.

If done after plowing, disking may be done any number of ways. Typically though, if you are simply trying to create a seedbed, then disking will have to be done in two steps. The first step is to break apart the furrow ridges left behind by plowing. Here a disk harrow with notched-edge disks will rapidly break apart the furrow ridges and begin leveling the field. If you are using a one-way disk, consider traveling on a center line that lines up with a furrow. A single-acting disk tends to pull dirt along the center line. After the first pass, a second pass with a double-acting disk harrow with smooth-edged disks is made. This path should be parallel to the plow furrows. Additional passes may be required. Harrowing (covered next) is the last step and can be combined with the last disking.

If you are disking in place of plowing, then a little more planning will have to go into it. If you are planting a drilled crop such as wheat, and the field is in good shape, then simply using a disk tiller may be the way to go. If the field needs extensive work, you may want to use a disk plow, which is a cross between a plow and a disk. It cuts and mixes soil like a disk, but the disk is large and the disk plow works at depths comparable to a plow. Then conventional disking can follow if need be.

Harrowing

Typically, in the narrowest use of the word, harrowing is the field method that provides the final groomed finish needed for the seedbed. In the broader and more common use of the term, harrowing is the act of creating a final groomed surface to any area being worked, whether it is a field, gravel road, paddock, or whatever. At the opposite end of the spectrum, there are some implements that have the word "harrow" in them, but they are not harrows in the strictest sense. The primary example of this is the disk harrow mentioned above: This implement provides a reasonably groomed surface ready for planting in some circumstances, but it is more correctly considered a disk implement that yields deeper tillage than harrows do. There are other implements such as bog harrows that fall under this category as well.

Depending on the type of harrow used, harrowing can be

Here are two antique tractors hard at work. The first is a 1952 Oliver 77 diesel mowing the hay field. Parker Yost is following behind on a 1954 Massey Harris Pacer tedding the hay.

done to scarify and aerate soils and pastures, level gravel roads, and groom any loose surface. There is nearly an endless variety of implements available to harrow with, and in fact not all will have the name harrow associated with them. A harrow can come in any shape or form, and, in fact, many are often homemade. Whether the harrow is simply a log dragged behind a disc, pasture rake used to scarify a pasture, or a chain harrow used to level the gravel of a driveway, harrowing does not deeply disturb the surface.

The most common harrows are the chain and rake harrows. The chain harrow works simply as a smoothing implement that provides final seed bed surface texture. It further breaks apart the soil and leaves a very uniform finish. A rake or pasture harrow has small teeth either solidly mounted or spring mounted that detach and "wound" the soil. This is used to aerate the soil and prepare it for the application of seed, fertilizer, and other additives. As I alluded to earlier, harrows can be homemade. Consider using sections of chain-link fence as a chain harrow, logs (pulled perpendicular to the centerline of travel) as a leveling harrow, or the old-fashioned homemade V harrows for seedbed preparation. The last pass made with a harrow should be perpendicular to the path you plan to take when planting. This makes the paths left by the planting equipment much easier to see and will help you keep your rows spaced properly.

Tedding hay is the act of turning the hay to improve drying times. This is a Daros brand tedder, which is a modern and well-respected brand. It is being pulled quite capably by an antique tractor.

Planting

There are three common methods of planting crops: drill, broadcast, and row crop planting. Examples of drilled crops include wheat, soybeans, and the like. Broadcast crops would be pasture grasses, legume winter cover crops, and annual flowers being raised for seed. In row crop planting, individual seeds are properly spaced within rows that are very regular and consistent. This method is used for corn, market vegetables, and some types of nursery stock. Both row crops and drilled crops are planted the same way and the seeding implements operate basically the same way. They trail behind the tractor and open a planting furrow at the right depth, deposit the seed at the correct interval, optionally apply fertilizer, and close the planting cut. The operation is complete and no other prep work is required. A good planter or drill will pay for itself in increased germination, healthy starts, and low seed waste. This is one area where a producer will buy the best he or she can afford. Many good brands of planters were available in the past when these tractors were new and some of those companies are still making planters today. Names such as Planet Jr. and Cole are good bets for high-quality antique planters, while makes such as John Deere (including the Van Brunt line that John Deere acquired and continued to sell with both the Deere name and the Van Brunt name) and Oliver made excellent drills. Good bets for modern planters are still Cole and Planet Jr., while good drills can be had from John Deere, Pasture Pleaser, Brillion, and Massey-Ferguson.

There are many differences between drills and planters. Drills usually plant very close together and plant shallow. Row crops are spaced more widely apart and are planted deeper. Drills will plant numerous rows at once, while most row crop planters only plant a few rows at a time. Drills accommodate for different seed sizes and types through the adjustment of sifting ledgers or holes in rollers that feed seed to the drop tubes. Row crop planters use different plates with different-sized holes to accommodate the different types of crops. Many drills will also, in a limited way, broadcast seed, while planters cannot.

Broadcast seeding is done through a distributor. These implements sling the seed in a pattern that evenly disperses it. These plates work very similarly to the push type of broadcast seeders and fertilize distributors that many of us have used on our lawns. The rate of application is controlled through adjusting a constriction at the bottom of the seed hopper that feeds the seed to the broadcast mechanism.

Post-Planting Activities

CULTIVATING

Cultivating is the act of keeping weeds at bay in your field and garden as well as a method for working in additives, and improving the structure of the soil. Cultivating is done only to row crop planted crops where there is enough room between rows for tractor tires and cultivators and there is enough light for weed growth. While cultivators are available for drilled crops, they are very infrequently used and are more common in Europe than anywhere else. In most drilled crops, weed growth is limited by close planting. Cultivation's popularity is waning in production farming, but it is still a mainstay for small gardens and truck farms. Cultivating was such an important part of farming in the past that many tractors were specifically designed for this task. Tractors such as the Farmall A and Cub were designed with offset drivetrains to enhance the operator's visibility during cultivating. Allis-Chalmers Gs were also an important cultivating tractor, with a rear-mounted engine for increased visibility. Also, cultivators were commonly sold with many antique tractors and are still available on the used equipment market. In fact, antique cultivators made specifically for a particular tractor far outperform modern three-point cultivators. This is one implement I would recommend using an antique version of, if possible.

The business end of the cultivators, the shoes (also called teeth or points), determine what type of cultivating is done. The variety is seemingly endless, but they all fall into three basic categories: weed destruction, listing, and tilth enhancing. The first refers to shoes that cut and overturn weeds. Most typical shovel types of cultivator shoes fall into this category. While I refer to the second category as listing cultivators, they are not necessarily used only on listing crops. This type of cultivator shoes cuts and moves the dirt toward the base of the crop. Shallow rooted crops such as corn especially benefit from this type of shoe. Field sweeps are the shoes commonly thought of as listing shoes. Tilth-improving shoes simply try to break apart soil crust, maybe work in an additive, and help cut and mulch weeds. Rotary hoes fall into this category. Often the application of a side dressing of fertilizer is incorporated into the cultivation step and is usually accomplished through the use of side-mounted fertilizer hoppers driven by the rear axle of the tractor.

SPRAYING

Another farm and ranch chore your antique tractor will be adept at is general purpose spraying, whether you're applying orchard sprays, added liquid fertilizer, or simply watering down a dusty road. Most sprayers are either gravity fed or PTO driven. This means any antique tractor with standard PTO, which is the majority, will be up to the task without any added equipment or retrofitted devices. The only exception to this rule are sprayers driven by hydraulic motors and sprayers that are not towed behind the tractor, but are supported by a three-point hitch. Fortunately, both of these types of sprayers are out of the ordinary and can be substituted with other types, or the tractor can be retrofitted with a PTO-driven hydraulic pump or a three-point hitch. The only thing to be aware of with sprayers is that the pumps require that the PTO be driven at the standard

---

Antique Tractor Tip

Clutch Tricks for PTO Use
Always remember to use an over-running clutch when using a PTO-powered implement with an antique tractor.

---

speed to maintain pressure. Higher speeds may damage the pumps, and slower speeds yield poor results. Newer tractors have the tachometers and instrumentation necessary to assure proper PTO speed, but most antique tractors do not. You may want to invest in an aftermarket tachometer to help judge proper PTO speed (engine-rpm to PTO-speed conversions should be available in the operator's manual for the tractor). While most antique tractors with a PTO will turn it at the standard speed at top or nearly top engine speed, many antique tractors are capable of turning PTOs at much higher speeds. Farmall Cubs not only run their PTO at a very high speed, they also turn it backward from standard.

Harvesting and Combining

Bringing in a grain crop is a complicated procedure that typically involves two steps, reaping and threshing (when not done by hand). Reaping and threshing have specific meanings that vary with the crop being discussed, but I will use them generally here. Reaping is the act of cutting the crop. If you

While cultivating beds and row crops, cultivating the soil behind the wheel to remove any soil compaction created by the weight of the tractor is advantageous in most soil conditions. These John Deere wheel scarifiers were designed to do just that.

hand-pick corn, you are not reaping since you leave the stalk standing. To automatically harvest corn, or most other crops for that matter, the plant must be reaped. Next, the loose crops are organized and packed for transport to threshing in a process called binding. Then the crop is threshed, or separated from its plant material.

Harvesting with hand implements such as scythes and sickles, and then hand threshing, may be possible if the grown area is small, but putting your antique tractor to the task is a lot quicker. And more fun! Unfortunately, harvesting equipment can be expensive, and many full-time

farmers with large farms even hire out harvesting for that reason.

Most antique harvesting implements are either reapers, threshers, or combines—a combination of the two. Antique combines are usually large threshers that have the appropriate cutting devices, or heads, attached to the front of them. These combines are either ground-wheel driven and were pulled by a team of horses or a tractor, or are PTO driven. Modern combines are self-propelled, air-conditioned, and handle the entire harvesting from cutting the crop to loading the trucks. Self-propelled combines became popular many, many years ago, so there are very few modern combines available today that can be powered and pulled by antique tractors. The antique tractor owner's best choice in harvesting equipment is to look for antique pull-behind combines and threshers and integrally mounted corn pickers. Do be aware, however, that their efficiency cannot compare to that of modern equipment. If you are planting for market, hiring out a local combine operator to harvest your field may actually be a better use of your money than investing in an antique thresher or combine.

Haying

Haying is probably one of the most common tasks for antique tractors, and they handle it exceptionally well. Many small farms would have no need for an antique tractor if it weren't for this chore. Although

younger antique tractors with live PTO (or at least those with two-stage or double clutches) and three-point hitches are best suited for the task, any antique tractor with a standard PTO unit can be used. Most haying chores do not require a large tractor, and in fact some of the tasks can be done with a small tractor. Mowing, fertilizer distribution, harrowing, and seeding can all be done with any antique tractor. Raking and conditioning can also be done with these machines, though using a larger tractor would be easier on the tractor and operator.

The one task in haying that requires a fairly large tractor is baling the hay. The smallest balers, which create small square bales and are used on reasonably flat fields, require a tractor with a minimum of 25 belt/PTO horsepower. The horsepower requirement increases to 30 if there are any hills to be hayed. Smaller tractors can be used if the baler has its own power source, but the weight of a baler with its own power source is significant, and tractors that weigh less than 3,000 pounds after ballasting are not recommended. Even then, the weight of the tractor and a baler with an engine may be too great for a smaller tractor's brakes. Consulting a local agricultural agent or experienced farmer or tractor mechanic can help you decide if your planned combination of baler and tractor is safe. Many operators, regardless of the types of balers, find that adding weight to the tractor helps to minimize the rocking motion created by the baler and helps reduce wear and tear on the tractor and operator.

Mowing

Mowing can be done with any number of implements, the most common of which is the sickle bar mower. Sickle bar mowers in good repair are quiet, reliable, efficient, and good for both the plant and the harvest, and leave the hay spread well for initial drying. In addition, small tractors with small fuel needs can power a large mower. Mowers made for your specific model, as well as modern three-point hitch varieties, are available. The downside to both types is that they are maintenance nightmares, prone to breakage, and considered slightly more dangerous than some of the other mowers.

Other types of mowers abound and all have their strong points. Rotary cutters have several advantages. They put the hay in windrows right from the start (not recommended for any humid climate) and are probably the safest among the different types of mowers. These cutters can be bought so they require only a drawbar and PTO and do not require a three-point hitch. The cutters can be modified to spread the hay out more evenly by removing the side of the deck of the rotary cutter, but then the machine becomes more dangerous to use, and the hay is still not as evenly distributed as one would like. Discbines are becoming very popular, as they cut and leave the hay evenly distributed like sickle bar mowers, and minimize the flying object danger of rotary cutters. Their downside is that they are of recent development,

so you must have a three-point hitch on your tractor. They are also large, requiring more tractor than the sickle bar mowers, and quite expensive.

MOWER/CONDITIONERS

Conditioning is one of the optional tasks in putting up hay. To condition the hay, most conditioners pinch or crush it between grooved rubber rollers, creating broken places in the hay that help to speed the drying time and allow for more even curing of the harvest. This process is also called crimping, so conditioners are sometimes called crimpers. These devices cut the time needed for field curing considerably, which in wet areas of the country can mean the difference between bales in the mow and ruined hay. Many conditioners are available with a sickle style mower, allowing you to mow and condition in one step. These mower-conditioners have the same advantages and drawbacks as sickle bar mowers, but have the added advantage of conditioning in the same pass. None of the other mower types can be integrated with a conditioner.

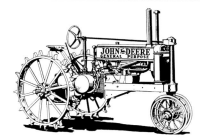

Raking and Tedding

To allow for even field curing, the hay must be occasionally mixed and turned. This process is called tedding the hay. Tedding is often optional, especially if the hay is conditioned. In fact, tedding is discouraged if you are baling a leafy or grain crop, such as alfalfa or soybeans, since tedding tends to knock off leaves and seed heads. The tedding is usually done with the same implement that you rake with, but the implement is set to throw the crop in the air. A perfect example of this is the star or pinwheel rakes. These rakes, when rotating in one direction, gather the hay and force it into windrows. When turning in the other direction, they ted the hay and leave the hay evenly distributed.

Raking the hay simply means that you are gathering it into long serpentine piles called windrows. These piles are then navigated with the baler to gather the hay into bales. Often, the hay is left to cure while in these windrows. Letting the hay cure additionally while it is in windrows is called sweating the hay. If this is

Antique Tractor Tip

Belly Belts

Many makes of belly-mounted mowing decks made for antique tractors require very long and expensive belts to power them. Be sure to frequently inspect the belts, repairing any cause of chafing, stretching, or unusual wear.

done, the hay is usually raked one more time, turning the windrow over. Then it is baled. Side-delivery rakes are the most common rakes used with antique tractors because they do not require three-point hitches. These rakes were common at the time these tractors were made and are less expensive than the star or pinwheel rakes. Side-delivery rakes do not ted the hay, however, so a separate implement is required for that purpose.

Baling

Baling the hay is straightforward, and following the baler manufacturer's instructions is all that is required. Balers are pulled by the drawbar and therefore have no special hitching implications. There are a couple of things to keep in mind, however. Antique tractors typically do not have a live PTO, that is, a PTO shaft that continues turning even when you push in the clutch. Oftentimes when baling hay you want to have the baler continue operating, but you want to stop forward travel. You can't push in the clutch

since that would stop the baler. To stop forward travel yet still have the baler operating, you must have the tractor in neutral. When you do this, however, be aware that the energy in the baler flywheel will continue to turn the tractor's rear wheels, via the PTO shaft, leaving you completely unable to stop even when the clutch is in! This problem can be cured by a device placed on the PTO shaft called an overrunning clutch, which I consider required equipment.

This problem of a non-live PTO shaft renders brakes all the more important, since stopping forward travel and leaving the tractor in neutral will more than likely happen on hills where the engine is having a difficult time handling the baler and the terrain. Avoid pulling a baler on any incline you think your brakes might not be able to handle. Trailing a hay wagon behind the tractor-baler may also overstress your brakes.

Another thing to keep in mind about hills and thick windrows is that antique tractors often do not have a first gear that is slow enough to bale in particularly thick or steep

conditions. The only cure for this is to try to rake your windrows so they are smaller, traverse the hill sideways, or use a tractor with more horsepower.

Miscellaneous Tasks

MOVING, DIGGING, AND LEVELING BULK MATERIAL

Very few tractors made before the mid-1950s had the live or remote hydraulics, or the three-point hitches needed to mount and power front loaders, fork lift attachments, carry-alls, backhoes, scrape blades, and snow plows. Retrofits are necessary to handle these implements, and they typically are not difficult to find. They can be costly, though, and some require quite a bit of effort to install. Unfortunately, there usually is no substitute if you want to perform these tasks. The only exceptions are the snow plows and scrape blades, as they were available as attachments for most models of tractors when they were new and therefore are available on the antique implement market.

Front loaders and backhoe attachments fall under the same category, and both add incredible functionality to your antique tractor. Unfortunately, both can require extensive modifications for installation. Loaders and backhoes are powered by hydraulic pumps that pump hydraulic fluid, under very high pressure, into rams, or hydraulic cylinders. The amount of hydraulic fluid and the pressure these rams require is beyond the capability of most antique tractor hydraulic pumps, if the tractor has one at all. To add insult to injury, often the hydraulics are not live; that is, you must have the clutch engaged before the pump will operate. This is particularly troublesome with a front loader. I operated a tractor with a Wagner front loader one time to move mulch. It was bothersome to have to be either in a gear, or in neutral with the clutch engaged. Those of us who have operated newer equipment, especially, become irritated when we are used to raising a bucket, pushing in the clutch, and shifting to reverse all at the same time.

Therefore, to use front loaders and backhoes, an auxiliary hydraulic pump usually has to be added. In most cases these pumps are added to the front of the machine and are powered off the crankshaft, or they are mounted to and driven by the PTO shaft (the PTO must be live or you haven't accomplished anything). These pumps, in addition to the valves that control the movement of the implements, operate the loader, and the hydraulic system of the antique tractor (if it has one) is completely bypassed. This has several advantages. Even if the flow rate and pressure of your antique tractor's hydraulic pump is sufficient, neither parameter is probably as high as it could be for effective, efficient operation; plus, having the second pump keeps the tractor's pump from wearing out. In addition, leaving the tractor hydraulics free for the lifting mechanism allows quick changes in the way the tractor is used. Just leave the front loader on and start disking. The hydraulics of the tractor are available for the lift and draft

Here is another task for antique and classic tractors: Powering grain augers to fill a silo. This Oliver, while not truly an antique, could be considered a classic as it is a 1969 model.

control. To mount a backhoe, your tractor will need a three-point hitch.

Front loaders are not too terribly hard to find, though they can be expensive if you do. Backhoe attachments are harder to find and just as expensive. Both are available for some models of antique tractor new from current manufacturers. Buying used makes sense, as units on the used equipment market are serviceable at a lower cost than a new one. Do be aware that many antique loaders and backhoes will require attention and tend to leak oil considerably, however. Be sure that the unit is in reasonable enough shape to operate properly. Rebuilding hydraulic components, rams, and control valves in particular can be expensive.

Loaders and backhoes place greater stresses on a tractor than many implements, which can lead to component failures and

dangerous situations. Older, frail, and failing hydraulic hoses, for example, can create a dangerous work environment. Check them over carefully and have a hydraulic mechanic help you if you're not sure of a component's condition. A front-end loader will strain the tractor's front end, increasing wear on the axles, spindles, and axle bearings; narrow front tractors are particularly susceptible. Front-end structural components on antique tractors have been known to break while using front loaders. Also be aware that there will probably be no protective screening to prevent injury from the loose material in the bucket. One last thing to remember is that antique tractors have a high center of gravity, and a front loader or backhoe bucket filled with dirt can radically affect stability.

Scrape blades, box blades, and pushing plows all make a great team with an antique tractor, even if your tractor has just a hand lift. I have used a homemade scrape blade mounted on the underbelly cultivator frame of a hand lift Farmall A; a three-point scrape with a Ford NAA (also called the Jubilee, which has live hydraulics but no down pressure); and a belly-mount scrape with a Massey-Harris Pacer (hydraulics with down pressure). I can say that the homemade scrape on the Farmall was the most effective for typical leveling and snow plowing. The hand lift had positive height-stop adjustments, three-way blade adjustments available with the turn of a wrench, and the precision that only a belly-mount blade can offer. The Pacer was second best only because of down pressure hydraulics: The hydraulics would drift over time, requiring height adjustment. The Ford NAA is about worthless, and can barely scrape a gravel road unless tremendous weight is added to the blade. So in other words, don't discount your antique tractor for this type of work if it only has a hand lift and no hydraulics or three-point hitch.

Front-mounted leveling plows were available for many types of tractors, and they do an excellent job with some types of work, especially snow plowing. Modern front-mounted snow plows and scrape blades are normally not available separately, but are usually available as part of a modern front loader retrofit for an antique tractor. You simply interchange the front loader bucket with a blade that will push loads. Pull-behind style road

graders, scrapes, and terracing machines were more common in the past than they are today (in fact, I doubt draft style road graders are still made). These machines did excellent jobs at leveling and grooming roads, building terraces, and performing soil erosion work. If you can find one of these machines in usable condition, you can turn out professional quality land and road finishes. While finding one and then finding repair parts would be problematic (but not ridiculously so), the job quality from these machines nearly makes the effort worthwhile. These are further examples of antique implements that perform the task much better than their modern counterparts.

POST HOLE DIGGING

Although antique post hole digging implements are available, most folks will find that modern post hole diggers are more readily available, cheaper, and easier to use. Most antique post hole diggers used the belt pulley of the tractor for power. Modern post hole diggers, also called augers, attach to the tractor via a three-point hitch and are powered by the tractor's PTO shaft. This implement is much easier to use with a tractor with down pressure-style hydraulics. Without it, you must rely on the weight of the auger to provide the bite needed to dig the hole. In all but the softest soils this is slow or impossible. If your tractor does have down pressure hydraulics, be careful that the auger doesn't screw into the ground so rapidly that it raises the front of

the tractor. This can happen when the auger catches on a root or rock while cleaning out the hole and less frequently while digging.

STATIONARY LOADS

Antique tractors are well suited for many stationary loads, and in fact the need to power a stationary load was often the reason farmers in the early 1900s bought their first tractor. Even steam tractors at the turn of the century performed stationary chores, such as powering a sawmill, thresher, or grain mill. Today, the number of stationary loads that tractors can power have multiplied to include generators, hydraulic pumps, irrigation equipment compressors, and others. If your antique tractor has a PTO shaft, it should be able to handle any stationary load you can purchase or devise, or at least do so with very little modification.

People often use antique tractors to power generators. PTO-powered generators are available at any farm or construction supply business and are priced quite reasonably compared to generators with their own power sources. To match a tractor to a generator, you need to convert between kilowatts and PTO horsepower. In a perfect world 1 horsepower is equal to 746 watts, or about 3/4 kilowatt. However, inefficiency and transmission losses will reduce the amount of energy actually generated. Factor in voltage drops, unequal loads, generator conversion efficiency, starting current, and old, worn-out tractor engines that are not capable of producing rated output. This leads us to de-rate our ability to generate a true 746 watts with each rated horsepower of the tractor; instead, let's use the figure 500 watts, or 1/2 kilowatt, per horsepower of rated output. Looking up the horsepower rating for your tractor in the Nebraska Tractor Tests, let's say you find that your tractor is rated to deliver 17.2 horsepower at the belt pulley or PTO. We then multiply that times 500 to get the number of watts your tractor will produce. In this example, the tractor will accommodate an 8,600-watt generator. As soon as your electrical requirements reach 8,600 watts, increasing the load will begin to max out your tractor. Eventually the tractor will stall if pushed beyond capacity.

Grinders, Choppers, Mills

Among the varied stationary loads for antique tractors are the class of loads that process forage and grain into food or livestock feeds. These loads were among the first applications for mechanical power sources, and antique tractors handle them well. The two main types still being used are small hobbyist mills and grinders and choppers of modern manufacture. These are made primarily for small farms, stables, and even small breweries. Modern units are often driven by a belt, powered either by an electric motor or the PTO, though some are driven by the PTO itself. If you make modifications to drive the implement with a belt from your PTO, be sure that the final operating speed of the grinder, after reductions and multiplication

Putting a gate valve on your implement's hydraulic hose prevents leakdowns while disconnected and allows you to attach the hose to the tractor with less effort if the hose coming from the implement is under pressure, such as when the weight of the implement is resting on the hydraulic ram.

through gearing and pulleys, is within the manufacturer's specifications. Antique mills, choppers, and grinders, while not exceedingly common, can be found and purchased at antique farm equipment shows. These units range from very poor "just pulled out of the storage shed" condition to newly restored units that should last a lifetime. The only problem with most of these units is that you will need to find parts for them eventually, and this can be difficult, if not impossible. Some parts may have to be custom-made. The stones in the various types of mills are difficult to find as well, and I recommend buying a unit with a complete set of stones. Hammer mills, or mills that process grain by beating the grain with a metal hammer against a metal anvil, do not require stones and are

preferable for this reason. Sizing a unit to your tractor, while not unimportant, is not critical. As long as the tractor has the power to drive the mill or chopper with some reserve available, you will be able to process grain and forage. Simply alter the rate at which you add the material to compensate for the power level of the tractor. Many of these old units are powered by belt pulley and not by a PTO shaft, and therefore a belt-pulley attachment on your tractor is preferable. Of course, using a PTO-to-belt converter (a small stationary device that takes PTO power input and turns a belt pulley) is an option, if need be.

The types of stationary loads available for your antique tractor seem endless and limited only by your ability to design, build, find, or purchase them. There are so many, in fact, I won't be able to mention them all. Following is a partial listing: hay elevators, grain augers, conveyors, and blowers. Belt-driven stationary loads, such as threshing fans and silo loaders, can be used with any antique tractor with a belt pulley. Stationary hydraulic systems, such as log splitters, hydraulic motors, and lifting systems can be driven from a PTO hydraulic pump. Winches and cable lifts can also be driven by your antique tractor's hydraulics, though some modification may be required. One of the most novel uses of antique tractors occurred on the shores of North Carolina in the 1930s, 1940s, and 1950s. When the Spanish mackerel were running near shore, rowing teams would row out to set the nets, and then tractors would pull the nets ashore. In the later years, winches were added to the tractors and the tractors remained stationary while the winches pulled in the catch.

POST REMOVAL

Pulling a dead load, such as a post, from the ground using the drawbar of the antique tractor is an exceedingly dangerous activity if the operator goes about it in the obvious way. There is a tool available, however, that nearly eliminates the risk as well as increases the effectiveness of the tractor. In fact the flipover hazard is so great that I recommend you never even try to pull a dead load from the ground without this tool. The tool has many names across the country. I have always called it a T-bar, but the most common name seems to be rocking post.

A rocking post consists of two pieces of very heavy, strong wood or two steel bars joined in a T shape. At the foot of the T and running perpendicular to the crosspiece that forms the top part is a notch large enough to accommodate a strong chain. Now the T is turned upside down. With the top part of the T on the ground and oriented so it is perpendicular to the line of travel, a chain is routed from the tractor's drawbar through the notch (and secured there via a bolt or pin), and down to the fence post or stump. This provides leverage and pulls the stump from the top. Obviously there are limits to the size of the post or stump you can pull using this method.

These bars have three attributes that make them required equipment: They increase the leverage applied to the post, pull the post from the direction that offers the least resistance, as well as alter the centerline of pulling to minimize the force responsible for a backward flip. The best thing about this device is that it typically can be site built rather than purchased.

The next best way to use an antique tractor to remove stump posts and stones is to outfit your tractor with a winch and use the tractor simply as a dead man to anchor the winch. Be sure the winch can be remotely operated for safety reasons. The third best way to remove a dead load from the ground using an antique tractor is to sit on the seat, use a cellular phone, and call a qualified heavy equipment firm to bring a backhoe or front loader to remove it. Rearward flipovers are most commonly caused

by trying to remove dead loads or moving them once they are free. This activity is no small matter and only experienced operators careful to control the conditions and stack the odds in their favor should attempt it.

REMOVING A STUCK TRACTOR FROM A FIELD

If you use a tractor long enough, you can be assured that eventually you will get the tractor in soil too soft to support it. Getting a tractor stuck is common, but what is uncommon is the presence of mind to realize that continuing to try to get the tractor out on its own power is futile and makes matters worse. After a few quick initial attempts using a higher gear than normal, such as second or third (less likely to spin the tires in these gears), give up. Continued attempts to rock or spin the vehicle out will only dig deeper holes and make it much more difficult to get the tractor out. Sinking into the ground deep enough to mire axle or final drive housings guarantees that only serious heavy equipment, such as a bulldozer, will be able to remove it.

There is only one sure method for removing a stuck tractor: hiring a recovery or heavy equipment company. These companies will have the tracklaying vehicles, winches, lifts, and so on to get the tractor out quickly and without damage. If you are being pushed by the weather to finish an important field chore, then hiring a company such as this can be money well spent. If you're not one to wave the white flag so easily, the following ideas can help you undo the predicament yourself.

Your best first attempt should be to try to back out. This approach uses reverse gear, which usually has a ratio that's less likely to spin the tires than first gear (though cleats on tractor tires are nowhere near as effective in reverse). Second, backing out lets you use the tracks you made into the area to get you back out. If an implement is attached, lift it in the air or remove it. Lifting it may help to increase traction. If the soil is so boggy that the extra weight makes matters worse, or if the implement can't be lifted, then try, with a winch or another vehicle, to move the implement out of the way so you can try to back out. Groom the tracks if need be. If this doesn't work, pulling it out with another vehicle is probably the second-best approach. This is especially effective if there happens to be any dry sure ground near the stuck tractor. If there isn't, then you should try to pull the tractor backward using another tractor or four-wheel-drive truck and a long chain. Pulling the tractor from the drawbar using its previous tracks is best; pulling it forward usually forces you to chain to the front axle, which isn't designed for heavy loads.

At this point, you either don't have a tow vehicle available or you have tried a tow vehicle but the tractor is hopelessly mired. The next step is to try to lift the rear of the tractor using a cribbing of wood and hydraulic jacks. Another method that is handy in areas where logs are plentiful is to use a block and tackle and a tripod of logs. The tripod, for

Hay mowers have evolved from the sickle bar mowers of your grandfather's day. The disc mower is common on North American farms. Its set of rotating discs are each edged with two small flail knives that cut the hay cleanly and without as many jams as sickle bar mowers.

This rear shot of a John Deere mower conditioner shows the conditioning part of the implement. These steel rollers are mated to hard rubber rollers (not seen here) and crimp and crush the hay or straw as it passes through the mower. This crimping action helps speed the forage's drying time.

safety, should be lashed with chain, which will in turn support the block and tackle. Once hoisted in the air enough to clear the mud, heavy timbers can be laid under the rear tires. Then, laying timbers behind the tractor railroad tie style, you can back the tractor out of the soft place on a road of timbers. Short of hiring a professional, this method is probably the safest and most reliable way to remove a tractor from a badly mired position.

The Role of Your Local Agricultural Agent

Part of most states' land grant university systems is a program of assistance to state citizens and landowners involved with agricultural work. This is called the Agriculture Extension Service. While they can and do perform all types of farm, garden, and turf consulting, such as helping city dwellers with landscaping questions, suburbanites with soil analysis, and youth with the 4-H program, their main mission is to provide those engaged in agricultural work with any assistance and consulting needed to be successful. In the final analysis, these agents represent your best resource for many of your agricultural questions: They know your area and its farming history, soil, and climate better than anyone else. Your agricultural extension agent will be able to tell you the times of year to plant, what seed varieties do best for your area, what pests must be guarded against, and what plowing implements and plowing methods to try. Your Agricultural

Extension Agent is a great resource that successful operators of antique tractors learn to rely on.

I hope you enjoyed this foray into antique tractor usage, and I hope you understand now the types of things you can and can't do with your antique tractor. I hope you also understand that a book of this scope can only hope to expose you to the ideas and give you a quick synopsis and some good tips and ideas. As you can probably tell, the antique tractor is nearly as adaptable and flexible to your use and circumstances as any modern tractor. The ways you can apply it to perform work are limited more by your imagination, time, and resources than by inherent inability on the tractor's part. It is important to realize, however, that some uses for your antique tractor are not appropriate. Antique tractors typically have high centers of gravity and are not as stable in some circumstances as modern tractors. Antique tractors typically do not have the safety systems in place that modern tractors do and therefore special respect has to be paid to accident avoidance. Certain applications are poor choices for tractors in general, and antique tractors are no exception. Uses such as logging, some industrial, and some roadside applications are better left to specialized machines. Common sense, a healthy respect for safety issues, and forethought will all serve you well wherever and however you use your antique tractor. You should strive to learn more and proceed carefully. Have fun and put that old iron to work!

Using Antique Tractors for a Small Farm or Business

Using antique tractors productively in business and farming is not a strange or novel idea: It makes sense and is done every day. What I hope to do in this chapter is help you understand the economic, management, and safety issues that arise when you consider using antique tractors in business or farming. The decision is not as simple as it would appear on the surface, and requires some serious reflection and consideration. Your final decision will have a lot to do with your tolerance for the repair, maintenance, and safety retrofits required of antique tractors. For example, organic farming usually is intolerant of farm equipment with engine, transmission, and hydraulic leaks, which are typical on antique tractors. Read on, enjoy, and I hope you can use antique tractors on your business or farm.

Umbrella shades are available with universal mounts, or as in the case of this umbrella, made especially for your tractor's model. Anyone planning on using an antique tractor as a work tractor might consider purchasing one at any farm supply store.

Farming and Capital Requirements

ECONOMICS 101

There are certain things about every enterprise, whether it is a business or farm,

One of the many tasks antique tractors perform is mowing large areas of lawn. Here a Farmall model B has a Woods belly mower. It does a nice job and covers a lot of ground very fast. I mowed for years with a Woods mower on a model A, a tractor with an offset drivetrain to allow the operator a more complete view of the ground under the tractor.

Gene Dotson takes care of 20 acres of soybeans with his Case tractors. That's part of the beauty of antique tractors. Unlike other collector items, like cars, that become less serviceable with age, antique tractors are both collectible and nostalgic.

Non-hydraulic grain wagons, such as this one, represent a chore many large antique tractors still perform. In fact, any implement or job that requires nothing more than brakes, an engine matched to the task, and a strong hitch are candidates for antique tractors.

that are undeniable truths. The first is that every business must take some type of input, add value to it, and then sell it for a price that is higher than the sum total of the costs. Adding value to the product is where we provide skills or processing to the inputs to transform them into something more valuable. Most processing in today's modern society is done by machine, and these machines are known as capital. Farming is no different and, in fact, is a very capital-intensive operation. Things like land, barns, tractors, and so on are all capital. Compare this to a barber shop, where the only capital is a couple of chairs and scissors. Barbers don't buy the input, and they sweep away the output. The barber shop consists solely (or nearly so) of inputs that add value. The biggest part of the barber's economic equation are his or her skills and knowledge. Compared to the barber shop, you can see why farming requires a great deal of money. In fact, capital requirements in farming are so tremendous that farmers are constantly looking for ways to minimize them, and that is why antique tractors are often used on farms. They can minimize capital expenditures without reducing, in any significant way, productive capacity of the farm.

CAPITAL AND LABOR: THE PENDULUM SWINGS

Before the days of equipment, the only inputs into the business of farming were skills, knowledge, and labor; capital expenditures were small. Machinery and other

capital now replace much manual and skilled labor; activities such as welding, mechanical repair, and other traditional skilled activities are virtually non-existent on today's farm. Using antique tractors on modern farms and businesses brings the need for some skilled labor back into the economic equation. This is because antique tractors, no matter what their condition is when you bring them to the farm, will require frequent maintenance and repairs.

To make economic sense, antique tractors used for work must be kept running and productive. This will require skills and knowledge on your part as well as a certain amount of expense. The skills are learnable, and the satisfaction and financial security that comes from being self-sufficient should be considerable. In essence, what you do when you buy an antique tractor is trade lower up-front capital acquisition costs for additional requirements in the form of knowledge, skills, and labor. In today's society with increased discretionary time, it is ironic most choose to spend this "leisure" time in the pursuit of further business activities, such as hobby farming. Trading this extra time, our skills, and knowledge for lower up-front acquisition costs is the key to making antique, and "near antique," tractors work for you. On the other hand, you will find antique tractors more of a toss-up financially if your discretionary time is not sufficient to allow working on and maintaining your antique tractors yourself.

Antiques and Productivity

Productivity will be diminished somewhat because these older machines are not quite as capable as the newer machines, plus they are more likely to have time down than newer machines. They also place additional requirements on your time to maintain and repair them. Other issues with antique tractors that take a bite out of productivity are nonstandard hitching systems, more expensive fuel (most antique tractors are gas engine tractors as opposed to diesel), and the higher cost of parts searching and acquisition.

Fortunately, much of this can be minimized. Many antique tractors came with three-point hitches, or can be retrofitted with them from available kits. You'll want a three-point hitch if you will be changing implements often. Antique tractors can also be relegated to special uses by keeping a specific implement attached to them all the time. Saving antique tractors for light-duty chores also will minimize maintenance and repairs. The costs and problems associated with finding parts can be minimized by buying only very common models of tractors or buying a couple of parts tractors to have on hand for when problems do arise.

FITTING OLD IRON INTO THE ECONOMIC EQUATION

You use antique tractors in many different ways. They can handle

A small mid-mounted sickle bar such as this 4-foot mower mounted on a Farmall Cub makes a handy road mower or a mower for small acreage.

When storing tractors outside or in unheated buildings, be sure to drain the water from the engine or use anti-freeze in the cooling water. Otherwise the water in the engine will freeze during the cold winter months and create cracks in the engine castings. This cylinder head had freeze cracks develop one winter and was repaired through welding. This repair held well but it is often difficult for even a professional welder to create leak-free, long-lasting welds because of the expansion and contraction of the metal as the tractor is used. It tends to re-open the cracks at the welds and leak again.

Wheel weights are often needed for a work tractor to increase traction, but they are also a nice addition to any restored collectible tractor. Here is an example of rear and front wheel weights for a Case tractor.

Many manufacturers have developed proprietary hitching and implement attachment designs over the years. This Minneapolis-Moline has a plow attached with a spring-loaded frame for absorbing hard shocks and rapid loading, reducing stress on equipment, and providing a small measure of operator comfort and additional control.

almost any drawbar load. Unique and uncommon jobs are an antique tractor's specialty. Some examples are using the belt pulley to power equipment (modern tractors don't have belt pulleys) and front loading. A usable antique tractor with a usable antique front loader is cheaper than buying a front loader for a modern tractor. Since many businesses and farms use a front loader infrequently, older equipment is a perfect idea.

Many people use antique tractors as secondary and backup tractors. For example, being able to have a tractor that follows behind your modern tractor for finish work makes you much more productive. An example here would be having an antique tractor pull a hay wagon while the modern tractor is busy operating the baler.

OLDER MODERN GEAR — NOT LIKE A USED CAR

The many uses for an antique tractor beg the question: Why isn't antique equipment used in other industries? Why aren't cab companies running antique cars and delivery businesses using antique trucks? The answer lies in the expected use of the equipment. Antique tractors were designed to be used constantly and used in the worst conditions imaginable. Running a piece of equipment 10 hours a day at 85 percent capacity or better through muddy fields and dusty farm yards requires that the equipment be overengineered in just about every way. In addition, it was expected that the tractor would be refurbished many times in its life, and just about all parts and surfaces that wear are replaceable or repairable. With the exception of large road tractors, cars and trucks were not designed to this standard of utility, nor were they designed to be repairable to the extent tractors were. In addition, cars and trucks often are involved in collisions, rendering them useless and diminishing the supply of used vehicles. If this weren't enough, cars and trucks are typically more highly collectible than tractors, driving the prices for the antiques much too high for use in modern business.

Some Issues to Consider

The primary problem with antique tractors in modern productive use is the fact they do not have modern safety systems. Devices to protect the operator in case of a rollover must be added to the antique tractor, and unfortunately this can be more difficult and costly than most people wish. Operator guards, such as the guard over a PTO shaft, were often nonexistent or seem to be always missing when you buy them. Fabricating or locating these guards can be time consuming,

This is a nice example of an after-market cab available for antique tractors. Hiniker made these cabs and much more for years and still sells aftermarket accessories, implements, and more.

hydraulics that can deliver the amounts and pressures that modern hydraulic implements require. The lack of an all-weather cab can make using antique tractors very uncomfortable. While none of these represent insurmountable problems, they do have to be addressed and should be before one purchases an antique tractor.

SIDELINES AND HOBBY FARMS

Probably the most compelling use for antique tractors is on hobby and part-time farms. Operators of these farms usually have limited funds for purchasing machinery, and limited use for it, and have the time to repair and maintain it. Hobby farms are also more likely to grow and cultivate and raise livestock in more traditional ways, which involve tasks suited well to antique tractors. An example would be plowing. While plowing has all but disappeared from modern agriculture, many hobby farms still use plowing as their main deep tillage method. To this day there is no better plow than the antique trailing plow. In fact, the prettiest plowing job I ever had the pleasure of watching was done by a very large 1920s Rumely Oil Pull Tractor with a trail-behind, eight-bottom plow gang. The field looked like a work of art in its depth and its precision. Another chore performed on hobby farms but not in modern agriculture is mechanical culti-vation. Again, a set of antique cultivators matched to an antique tractor by the manufacturer are very hard to beat.

For the northern plains, cabs were almost a necessity for some row crop work, such as late wheat plantings. This aftermarket cab on this John Deere was fully sealed and fairly roomy for its time. A closer examination of the photo will reveal removable side curtains.

costly, or both. One other safety issue is noise related. Antique tractors often have no muffler, and adding one is required to meet OSHA guidelines on noise levels. Fortunately, this is usually fairly easy and inexpensive.

There are some technical issues to address with antique tractors also. Since most of them use gas-oline, there is the problem of now having to carry two different fuels: gas for the antiques and diesel for the modern tractors. PTO shafts on some antique tractors may be nonstandard in size, speed of rotation, or direction of rotation. Some antique tractors have hy-draulic systems that allow draft control; others don't. Most antique tractors do not have remote

In short, antique tractors can make excellent financial sense in modern agricultural settings if you are aware of the trade-offs they require and are willing to make them. While they save tremendous money up front in acquisition costs, they only make strong financial sense if you are willing to do the repair and restoration work on these tractors yourself. The large professional farm will find that antique tractors are best reserved for niche roles and as backup tractors, while smaller farms and hobby farms will find they are excellent choices as primary production tractors. Just remember to retrofit the tractors with safety systems and be ready to make the few changes antique tractors may require to meet your needs.

CHAPTER 10
Tools and Shop Procedures

Here is an excellent example of cribbing a tractor. Cribbing means to support a tractor using a lattice work of timbers. This sort of support is generally more stable and safer than jacks if you expect forces to be placed on the tractor that may move it slightly, such as when you remove an engine or when the surface is not hard, such as gravel or dirt, where you are working. On hard surfaces, especially where only one part of the tractor might be raised, such as a rear wheel, a jack is sufficient.

Antique tractor maintenance doesn't require a complete mechanics' workshop, but it does require heavy-duty tools. The first consideration when shopping for tools is quality. Cheap tools, or those of "homeowner" quality, are nearly useless with antique tractors. Either they will fail to do the job, destroy parts in the process, or simply will not last past a few uses. You'll need automotive-quality tools or better. These are typically well-engineered tools manufactured by a name-brand company. Brands of this quality include Craftsman, NAPA, KD, and others. These tools are the bread and butter of antique tractor repair and can be used to perform most maintenance tasks. Occasionally, however, you will run across the need for industrial or professional quality tools. Bearing pullers (any puller really), taps and dies, presses, impact wrench sockets, and the like all should be bought from a tool line considered to be heavy-duty or professional level. Tool manufacturers such as Snap-On, Mac, Proto, and others fall into this category. In addition, many tool

Jack stands are a must whenever the wheels come off an antique tractor. Here are two different styles of jacks. The two important parameters are R and R: reach and rating. Many jacks aren't designed for the high ground clearance of tractors and often won't reach to an axle or frame bottom of antique tractors, so be sure you get one high enough. Even then you often have to build a small wooden crib to place the jack to get the reach you need. The second R, rating, is the weight load it is rated for. Most small tractors can be handled by two-ton jacks. Otherwise use four-, six-, or eight-ton jacks, depending on the weight of your tractor.

If you have decided to restore an antique tractor, here is what I would consider a minimal selection of air tools you might need to acquire during the course of the project. Shown here is an air chisel, an impact wrench with a standard anvil (the part the socket mates to), and one that swivels. There is also a grinder, sander, and a gravity-fed HVLP paint gun.

manufacturers make two different quality lines: a moderately priced automotive quality and a heavy-duty industrial line.

There are a few tools that are more or less essential for tackling routine repairs and maintenance. First on the list is an automotive-quality drive socket set with about 15 to 25 sockets in it. The socket sizes should range from 3/8 inch to as far past 1 inch as you can find. I have a set that goes from 5/32 inch to 1 1/2 inch in 1/32-inch increments, and I find this to be a very complete set. The few times I needed a socket over 1 1/2 inches I have been able to rent it for just a few dollars or borrow it from a friend. You will need one socket wrench called a breaker bar (a long-handled socket wrench that does not ratchet). This tool is very important in removing the stubborn bolts you are bound to run across. This tool can suffice as your only socket wrench, but for convenience, acquire a ratcheting socket wrench, too, if you can afford it.

The next set of tools you should purchase is a set of ordinary combination wrenches. These wrenches should have sizes that range from 5/16 inches to 1 inch or more. You should have a 4-inch adjustable wrench and an 8-inch adjustable wrench. The toolbox should then be completed with an assortment of typical common hand tools such as regular, needle nose, and locking pliers (for example, Vise-Grips brand), regular and Phillips head screwdrivers of various sizes, and so on. To complete your toolbox, you will need a grease gun.

If you are going to use air tools, an air compressor is a must-have in your shop. Here is a single-stage air compressor with approximately 60 gallons of air storage. While a two-stage unit would be nice, their prices are often outside of home and farm budgets. In this case, a single-stage is sufficient with enough air storage and reasonable expectations as to how much air is available for applications, like sandblasting, that use a lot of air. Be sure to get a heavy-duty unit with cast-iron cylinders. Like many tools, you may be able to find used, older air compressors of higher overall quality than the modern ones available in the big-box retail stores. If you do purchase a used compressor, be sure you select one that has been lightly used or recently rebuilt, and that the tank is in good shape.

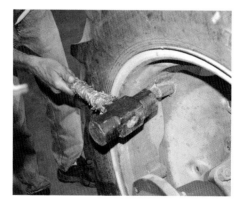

Removing stuck fasteners is a constant challenge on old tractors. Loosening the other side of the fastener, even just a little bit, and soaking it with oil often does the trick. A solid strike from a heavy sledge is occasionally what it takes to finish the job. Be sure to replace any fastener you feel is damaged.

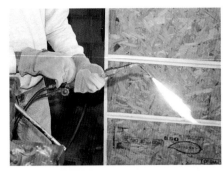

A gas welding outfit makes sense for anyone planning on restoring or repairing farm equipment. Here Steve Sewell outlines procedures for lighting a torch and welding. His torch is outfitted with a welding tip. Fuel and a little oxygen are fed to the tip using the torch's valves. Never try to adjust the flame by using the tank's valves. Using a striker, the flame is lit and the fuel is cut back until the flame burns bright and hot but with less intensity. Then the oxygen is adjusted until a strong, linear blue flame is achieved. Additional adjustments to the fuel and oxygen might need to be made until you get a flame that resembles the one he has here. Base metal composition and thickness might require additional adjustments. Steve is welding two pieces of bar stock at a 90 degree angle to make a temporary brace. To weld this, the joint is heated and the metal fill rod is placed in the molten pool as that pool begins to develop, filling the joint and binding the two metal pieces.

Another use of a gas welding outfit is to heat parts. This often makes disassembly easier, especially fasteners that are stuck. With a heating tip attached, Steve Sewell lights the tip similarly to a gas tip, but as you can see here, he is looking for a more aggressive flame. In the last picture he is heating a cast housing with a stuck spindle inside. The housing is heated to red hot. The differential thermal expansion of the housing thereby creates the space needed to easily drive the spindle out.

Lifting larger antique tractors and other heavy parts and assemblies require jacks with plenty of power, stability, and workmanship. Here is an air-powered hydraulic jack that is operated by compressed air rather than manual operation. The rate of lift is controlled by a trigger on the handle, and this air pressure replaces manual operation. The release of air lowers the jack. As with any jack, this is just for lifting. Once any object is off the ground, do not work under the load until you place stands, cribs, and braces under the object being lifted.

Electric arc welders are handy, but electric welders powered by a gasoline motor are even handier. These can be wheeled outside for a field repair or can be placed outside the shop door while running the leads into the shop for welding.

There are a few other miscellaneous tools that can come in handy. The majority of running problems affecting antique tractor engines involve the ignition or electrical systems. Tools such as feeler gauges, a multimeter, and tachometer/dwell meter will be indispensable in determining the causes of such problems. Sometimes precision measuring tools, like a spring scale or calipers, may be necessary for tasks such as adjusting the play of a bearing assembly. The last category is "bottom of the box" tools. These tools, which seem to work their way to the bottom of the toolbox, consist of punches, drifts, cold chisels, gasket scrapers, hacksaw blades, and the like. Most of these tools can be acquired when needed, though some are inexpensive enough that you should pick them up when you have the opportunity.

If you don't want to splurge on new tools, you don't need to. Unless you live many miles from a good hardware store, you can acquire them as needed. And sometimes used, high-quality tools can be found at auctions, estate sales, or business liquidation sales for very reasonable prices. Used professional-quality tools are better tools than brand-new, lower-quality tools, and the used tools are often cheaper. The best socket set I own I bought at an auction for $25; buying it new would have cost over $100. Next time you have the want ads in hand, see if anyone's unloading some good tools.

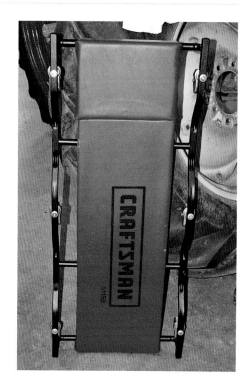

A creeper is a nice tool to have in your shop if you have concrete floors. While the high clearance of many tractors might lead you to believe creepers are almost useless, remember that many implements are low to the ground. It is also handy to use with your vehicles. There are many bull gear case drain plugs and hitch frames on tractors that are low to the ground.

A metal cutting bandsaw will be necessary if you plan on fabricating your own repairs and improvements to your tractors and implements. This unit operates much like a wood cutting band saw, but operates horizontally to allow the weight of the machine to create the force necessary for cutting. They are often operated within some type of a system to circulate cooling fluid on the cut while the saw is operating. This unit also has a small vice for holding the work in place.

This is the tried and true Lincoln electric AC arc welder. These units have been widely used for years in the farm shop. These units are also available as both AC and DC. DC mode welding typically gives smoother welds and better penetration for any given circumstance. However, for our needs, both AC and DC mode welding are strong enough.

This gas welding outfit shows everything you need here. Since the bottles need to stand upright, some sort of cart is necessary. Whether you fabricate or buy a cart, make sure the cart has large steel wheels. These are necessary for carrying the outfit over rough ground and gravel when repairs have to be done outdoors.

Keeping a supply of nuts, bolts, pins, clips, and other miscellaneous small hardware items is a real time saver while working on a project, but storing them in coffee cans is not the way to stay organized. Here is a nice stack of drawer chests that will keep your supplies close at hand and easy to find.

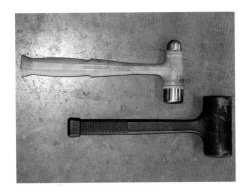

Here are two examples of mallets that are useful additions to your toolbox. Both are dead blow mallets. Dead blow hammers and mallets are made with shot or sand in a hollow cavity in the head of the tool to even out the force of the impact and enable a blow to be delivered while minimizing bounce-back and the risk of marring the part you are striking. The larger mallet here is also available in solid rubber for striking parts made of softer materials.

Field Tools

In addition to the tools you will use in the shop, there are a few other tools you will need to assist you in the field. Some implements require specialized tools that should be kept on the tractor or with the implements. You'll need a rivet punch and a ball-peen hammer for any equipment that uses peened rivets, such as a sickle bar mower. Leverage bars and pipes are handy when moving and positioning implements. A small and inexpensive set of hand tools kept on the tractor is a real time saver when the tractor requires attention far from your shop. You should also consider stowing a second grease gun on the tractor. Implements just never seem to make their way to the shop for maintenance, but you can add some needed lubrication if you keep a grease gun handy on the tractor or in the implement shed. Be sure not to overlook the need for other tools based on your own set of circumstances and implements.

Unsticking Stuck Parts

I can promise you that if you spend enough time around antique tractors, you will one day run across a bolt, screw, or nut that will not come off. Even after carefully trying

When sandblasting, you have a wide variety of sandblasting material, or media, available to you. On the left is a highly aggressive media made from coal slag designed to cut fast and remove a lot of material. This is great to use with highly rusted, heavy castings. In the middle are chopped nut shells (usually pecan), which is a softer material designed to be used on composite materials and thin metals. On the right is the more widely used sand. Also shown is a bag of crushed corn cob, which is another gentle abrasive that is safer to use around engines than other abrasives. Each type has its own place and use, but I recommend you start with sand (except the engine where you would use crushed corn or nut shells), and then move to another material only if you are not getting satisfactory results.

your best using all your might, it just won't budge. If you are like most, you will then try cussing, which helps you but leaves the bolt unimpressed. Unfortunately, in the rashness that stubborn bolts provoke, most of us then create extra work for ourselves by rounding off the bolt or nut, or grinding out the screw slots. This is a case where the typical approach to a problem isn't the best approach. Three approaches that do better with stuck fasteners than the wreck and rant method are using a different tool, using patience and oil, and drilling and tapping. (There are some further tricks beyond these three I will discuss later. They take more time, money, or effort.) Sometimes a properly sized tool, or one of different design, may remove the fastener. I typically try a different tool first, making sure I don't use too much force. A hand-held impact wrench often removes screws and small bolts better than a typical screwdriver or wrench. If I'm confident the tool has a tight grasp on the bolt or nut and is unlikely to strip it, a breaker bar or long pipe can provide enough leverage to free it up. There is a possibility of breaking the wrench this way: Never use a handle extension, such as a pipe, with a socket wrench.

Fasteners usually become stuck because of the formation of rust, corrosion, or galling between the fastener and parent material. By alternately applying heat and penetrating oil, you can often free up a seized fastener. To use this

This arbor press is one of the larger bench top presses available. Most presses of this size can handle much of your pressing needs.

This is a large floor model arbor press capable of pressing apart virtually any assembly you would run across. I recommend getting a floor model before a smaller bench model, as the larger models can often be set up to press apart small assemblies while the opposite isn't true. Smaller bench models won't have the power or structural strength to press apart large assemblies.

While a specialized abrasive is available for sandblasting, it typically doesn't make sense for the hobbyist to bother with the expense. Here the sandblasting pot is being loaded with a masonry grade of washed fine sand. As long as the material is washed and reasonably consistent in size, it will work for tractor restoration uses. Because it's inexpensive, you also don't have to bother with collecting for reuse, which is something you should do with expensive specialized abrasives.

method, alternatively apply oil, heat, occasional blows from a hammer, and patience. Never apply oil to red-hot metal or apply an open flame to pools of penetrating oils. Penetrating oil may work great or not at all. Apply some liberally at first and allow it to sit as suggested by the manufacturer. To apply heat, use a plumber's torch (often greater heat sources such as an oxygen-acetylene torch are needed) and direct the heat to the fastener itself. It may take from several minutes up to several days to

loosen an especially stubborn fastener. Occasionally tapping the fastener with a hammer will help, but be sure you don't distort it. Hitting the head of a bolt squarely is much less likely to do damage than hitting the threaded end. Unfortunately, the heat and oil method can take too long to work.

The last method to remove a stuck fastener—destroying it—is a last resort. Destroying the fastener works every time but it's tedious and time consuming, can create collateral damage, and in some circumstances is not an option. Drilling out the fastener is often the best way to destroy it. The trick is to drill out only what is necessary to remove the fastener without damaging anything else. This might mean drilling and then splitting off a nut, or drilling out a bolt or screw. When doing the latter, be sure not to damage the threads of the parent material. Using this method requires a high degree of precision on your part. If possible, try to remove the assembly the fastener is a part of and place this in the vise of a drill press. This is especially helpful if the fastener is hardened (heat treated, for instance), as hardened fasteners are very tedious and difficult to drill out while on the tractor.

To begin, whether you are at the drill press or the tractor, you first create a punch mark on the top exact center of the fastener. This will help you drill the fastener out straight and true. Start with a small bit and try to drill a hole in the bolt or screw that exactly follows the center axis of the fastener. This first

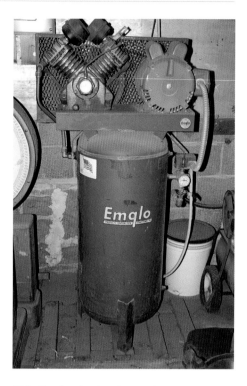

This Emglo single stage air compressor is a solid, reliable unit that would deliver all the air anyone considering antique tractor restoration would need. The vertical configuration is a space saver.

hole is important, as it will create the path for the remaining steps. After this first hole, drill with successively larger bits through the same hole, until your hole is nearly as wide as the bolt, being careful not to damage the threads of the parent material by overdrilling. At this point, you can then either use a tap to clean out the threads, or often you can simply turn and unscrew the remnants of the fastener, since the act of drilling out most of the material of the fastener usually frees it up.

There are some advanced tricks for removing fasteners. One trick is to weld a nut on top of the damaged head of the bolt or screw. This is done with any type of welding machine, but MIG welding machines usually are easiest to use. Some mechanics even will grind off the damaged head first, if possible. Then simply tack weld the inside of the nut to the top of the remnant of the bolt or screw. Use a nut whose inside diameter matches that of the bolt (e.g., use a 1/2-inch nut to remove a 1/2-inch bolt). This often works because the excessive heat caused by the welding will help to unfreeze the bolt, and the new, clean head gives you good surfaces for a wrench.

There are other fasteners, collars, rings, and the like that will be quite stubborn from time to time and a few tricks will help you out here as well. Cotter pins are soft and can be easily drilled out if they break off. Occasionally, internal and external lock rings rust into place. These are problematic since very little protrudes from the groove with which to work. To make matters worse, I have found that rings frozen into the grooves invariably are brittle and break at any attempt to remove them. Careful tapping with a hammer, using oil and heat (if possible and safe) and rocking with proper locking-ring pliers will sometimes do the trick. Patience and pene-trating oil are your best tools since force usually just shears off the ring even with the top of the groove, creating a groove nearly impossible to clean out. Muriatic acid, if carefully applied and quickly rinsed, may help loosen the rust that is keeping the ring in the groove. As a last resort, a machinist may have to remove the ring for you. Collars, slip

Here is a slightly larger, commercial version of a sandblasting pot. While sandblasting pots are incredibly useful, they require an air compressor that delivers a tremendous amount of air. A pot like this, which has a large nozzle size, probably requires a two-stage compressor. Be sure you have the air flow from your compressor to support the sandblaster you are considering before buying it.

tabs, woodruff keys, and so on all become stuck in place from time to time. Often force, such as placing the assembly in a hydraulic press, will remove collars and similar parts. Woodruff keys are fairly soft, and simply digging them out of the groove with a small hardened chisel will do the trick.

The successful removal of a frozen part starts with realizing as early as possible that conventional removal won't work, then using patience, proper tools, and a little common sense and ingenuity to find a way that will get the job done without frustration or damage to the equipment.

Cleanliness

One of the more important shop rules when maintaining antique tractors (or any piece of equipment),

A quality sandblaster, such as this one, is the right compromise between size, quality, and portability for the typical farm and home shop.

Parts can often be repaired rather than replaced. This coolant outlet elbow has developed a rust hole. Rather than go through the expense and time required for a replacement, it would be much smarter to repair it. The best repair in this circumstance is brazing. Brazing is the act of joining two pieces of (or closing holes in) cast iron using a copper bronze alloy. This engine was sandblasted, so before you paint and while the surface is clean and dry, braze the hole shut. First, file the hole back to solid metal. Then use a gas torch using a small welding tip to heat (top left) the surface surrounding the seam or hole.

Here Lou Spiegelberg of Spiegelberg Restorations lights his torch, adjusts the flame, and after heating the surface thoroughly melts brazing material from the brazing rod around the edge of the hole and slowly closes the hole a little at a time. At the end, you have a solid layer of brazing material (top right, bottom left) closing the hole (bottom right). Simply dress up the brazing with an angle grinder to smooth it up and you're done. This repair will last the lifetime of the tractor.

is cleanliness. Getting dirt and contaminants in the assemblies and systems of antique tractors is a bad thing and a very common cause of catastrophic system failures. Hydraulic systems in particular are susceptible to damage from dirt and water. Engines require extensive air-cleaning systems (which are often bypassed on antique tractors,

I might add) and transmissions are vulnerable to damage from grit. Whether you are greasing a fitting, changing the oil, working on wheel bearings, or rebuilding an engine, cleanliness is of the utmost importance. Accomplishing this is easy, and incorporating cleanliness into your maintenance routine will come naturally if you simply

make it a priority. Things like laundering rags, wet mopping shop floors instead of sweeping them or using an air compressor to blow out the dirt, keeping high-precision parts covered during storage, and wiping off an assembly before you disassemble it will contribute to keeping important and expensive systems clean.

Lifting, Cribbing, and Shuttling

The last general shop issue we will concern ourselves with is lifting the tractor safely, securing it off the ground, and shuttling it when it isn't working. Lifting the tractor is serious business, and safety should be everyone's top priority. This isn't like jacking up a car, especially if the lifting has to occur in the field. The two primary concerns surrounding lifting a tractor are instability caused by the unusual front-end configurations and the sheer weight of the machinery. Another variable is lifting conditions, since often the lifting must be done while the tractor is out in the field. Lifting the tractor requires heavy-duty lifting equipment, which, here again, you may want to consider renting. I have found that 3- to 8-ton bottle or floor jacks are the best types of lifting equipment. Three-ton jacks should be used with the smaller tractors and the 6- to 8-ton jacks should be used with the larger tractors. I use the floor jack when I am on concrete and I use bottle jacks in the field. Using bottle jacks in the field requires laying down 6x6-inch blocks of wood as a platform for the jack. These foundation blocks must be placed level and on dirt only. Placing the bottle jacks on rocks is dangerous! The rocks can split or teeter unpredictably, causing the tractor to fall.

The first step in lifting a tractor is securing the front end. If the tractor is a narrow-front tractor, then you cannot lift either rear wheel very high, since the tractor may tip. At no time should any tractor have the entire rear end lifted at once from a single lifting point. If the entire rear end must be lifted, use two different lifting jacks. Lifting from a single point will cause the tractor to tip and fall. Even most wide front-end tractors will be unstable, since most front axles are the floating type that pivot on one point. To secure the front end of a wide front-end tractor, fasten blocks of wood between the front axle and the structural member of the tractor. These should be fastened in such a way that they will not work loose. If the tractor has a narrow front, then bolting outriggers to the front will help to stabilize the front end.

Cribbing means placing strong, large wooden blocks, log cabin style, as structural supports under the tractor to hold it off the ground while work is done on it. Cribbing is necessary for replacing and repairing wheels and tires, engines, and other major systems. The material used to crib the tractor is almost always oak wood blocks that are 6x6 inches or 8x8 inches and cut to handy lengths. I find 2-foot cribbing sections to be about right. For smaller tractors, 4x4 pine blocks are sufficient. Wood is successful in this application because it is strong, inexpensive, and gentle to the tractor; it conforms to slightly uneven surfaces and has a certain amount of friction between pieces of cribbing and between the tractor and the cribbing. This makes it safer in circumstances where things are not perfectly level. The idea with cribbing is to set as much of it in place as possible before lifting

A bench grinder is almost a requirement in a shop. From cleaning parts with the wire wheel to sharpening a punch, smoothing a weld, or reshaping screwdrivers, it's hard to live without one. Fortunately they are widely available for a reasonable cost.

the tractor. All you have to do is lift the tractor, and then add just a quick block or two to complete the cribbing. That way, you are not under the tractor setting up a cribbing block while the tractor is up on jacks, which is very dangerous. Also, setting up the cribbing ahead of time forces you to have a safe and complete plan in place before lifting the tractor.

Shuttling a malfunctioning tractor is often necessary. Moving the tractor from one location to another, if the distance is appreciable, should be done on a trailer, if possible. Trailers provide the predictable loading

Antique Tractor Tip

Stay Warm
When designing heating systems for your shop, be sure to take the presence of flammable fumes and liquids into account.

When replacing bolts, be sure to use the correct grade (often referred to as hardness) of bolt. The three radial slashes seen here on these newer bolts indicate a grade 5 (add two to the number of slashes) bolt. Always replace the bolt with the same grade, but if the original is missing, you can take a look at the other bolts used in the same application. If the missing bolt was the only one needed for a specific application, refer to a parts manual for your tractor. Most parts manuals list and specify each bolt, especially if the strength of the bolt is important.

This can be home designed, but special care must be used when selecting wheels for the cribbing. In addition, the cribbing must be secured together: Never move a tractor supported by loose cribbing. Another instance where the tractor must be shuttled is when it is necessary to move large parts of the tractor. For example, when working on a clutch, it usually becomes necessary to "split" the tractor. This means the tractor is unbolted and literally split in half. Shuttling a heavy front end with an engine is difficult. Typically it is done with a hoist that rolls on wheels. This hoist supports the rear of the engine while the front is supported by the front wheels. Again, the pivoting feature of many floating front axles must be disabled through blocking before splitting a tractor! Narrow front-end tractors must be stabilized with outriggers, or the tractor must have the whole front end disassembled before splitting the tractor. In this case, the hoist carries the entire weight off the front end.

and unloading that leads to safe transport. See the earlier chapter on trailering. In other circumstances, where the distance is short, or trailering isn't feasible, then towing the tractor is probably your next best option. While this activity in and of itself isn't dangerous, it is easy to get into circumstances while towing that are dangerous. It can be dangerous to tow a tractor on a public road, through steep or wooded areas, and it can be dangerous if care isn't used to hitch the tractor and the tow vehicle properly. Most tractor front pedestals and axles were not meant for the shock loads of

towing, and hooking the chain up to many of these front parts will lead to failure of the front end, causing an accident. Be very careful to hitch the towing chain to a stable, solid part of the disabled tractor. Other things to keep in mind are to never exceed a few miles per hour and never tow a tractor whose brakes are not working properly.

Once in the shop, moving and shuttling the tractor can be difficult. Occasionally, either the entire tractor or a large part of it must be moved. Once the wheels are off, the cribbing system that is supporting the tractor must be movable.

Maintaining an antique tractor is easy, enjoyable for most folks, and within everyone's grasp to learn. Frequent, thorough maintenance will help ensure that your tractor remains reliable and maintains its value. Fortunately, the requirements for tools and supplies for maintenance are not too extravagant, and some creative purchasing on your part will help to minimize their cost. Remember to be safe, and remember problems such as stuck fasteners will crop up. Working through this with the steps outlined above will help to get you back in the field.

CHAPTER 11

Maintenance

Maintenance is key to keeping your antique tractor running as it should. The procedures are relatively simple and the tools required are not terribly extensive, but old tractors require more attention than the appliance-like machines built today. You may even find performing routine adjustments and lubrications to be extremely rewarding. However you feel about maintenance, read on to discover the simple maintenance you'll need to do and the tools you'll need to do it.

Goals of Maintenance

The primary goals of any equipment maintenance are increased longevity, improved productivity, and continued assurance of the fitness of a machine for its purpose. Regular equipment maintenance increases longevity by preventing wear, which in turn keeps precision mechanical parts aligned, synchronized, and within tolerances. Maintenance improves productivity by assuring the machine's availability and helping to reveal impending mechanical repair issues early while they

The best place to buy nonmaintenance parts, such as whole engine blocks or wheels, is usually a large consignment auction. Here is Don King's spring consignment auction in Columbia City, Indiana. This is just a small part of the auction.

Oiling and greasing seems like a continuous task with old farm equipment. Here is a selection of oilers, all with their own design attributes to recommend them for a given circumstance. They all have one thing in common: They are at least 20 years old. Modern oilers are typically of very poor quality. I recommend picking up these older, better quality oilers from auctions and antique shops.

are often easier and less costly to address. A well-maintained machine has all of its mechanical and safety systems in place and operational, increasing the machine's fitness for field, show, or pulling track. While

this all seems self-explanatory, these long-term goals are often forgotten or dismissed during one's short-term focus on day-to-day activities. Keeping those goals in mind will help to clarify what maintenance is important and what procedures can safely be postponed, if need be, as well as helping answer other questions as they arise.

Maintenance is the reason many of these machines are around today to be called antique. Maintenance was more of a formalized procedure in years past. Because so many of us have become accustomed to the modern technology and consumer patterns that brought us "greased for life" bearings and automobile leases, we often are complacent and less ritualized about maintenance simply because it isn't needed or encouraged on much of today's equipment. You will quickly find, however, that heavy equipment, even modern heavy equipment, requires routine maintenance. Setting up a schedule, developing an approach, and making maintenance a habit will make a large difference in your equipment's mechanical fitness, safety, and longevity. Some of the maintenance discussions that

follow will involve how you will store equipment; some will involve choosing lubrication products and other disposable supplies; and others will involve the schedule with which you will renew or adjust parts and systems. Some maintenance procedures will be as simple as greasing a fitting. Others, such as timing the engine or adjusting the valvetrain, will be more involved. None are outside the scope or ability of anyone who wants to learn and try. I encourage all owners to do their own maintenance: There is no better way to know your machine.

Of course there may be some maintenance jobs that you find easier to hire out, and that's okay too. As long as the work gets done, your tractor will be ready to perform smoothly and reliably whenever you need it.

GENERAL CARE AND SCHEDULING
While antique tractors are rugged and certainly don't need to be pampered, there are certain basic things you should do to ensure consistent performance and longevity. First, store the tractor indoors. While occasional out-of-doors storage won't hurt it, leaving the tractor unprotected regularly will lead to weather damage to the sheet metal, tires, hoses, wiring, hydraulics system, and other components. The storage building doesn't need to be elaborate; any shed will do. Another rule of thumb in general care is to inspect the tractor thoroughly before each use. Inspect the tractor well enough to identify loose bolts, tire damage, cracked or leaking hoses,

Antique Tractor Tip

Toolbox Tactics
Plan on keeping a grease gun and a few basic tools in the tractor's toolbox. If your tractor doesn't have a toolbox, making one is easy enough and it can usually be mounted in a nondamaging, nonpermanent way.

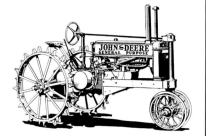

Suggested Schedule of Maintenance

	HEAVY USE (More than 100 hours a year)	LIGHT USE (Less than 100 hours a year)
Oil change	Every 100 hours or at least every 6 months	Every 6 months
Chassis lubrication high-speed joints	Should be lubed every use, (spindles on mowing decks, etc.) otherwise once a week	Once a month
Air cleaner (wet oil bath type)	Once a week (daily in dusty or chaff-filled environments)	Every oil change
Air cleaner (dry paper type)	Once a month (once a week in particularly dusty or chaff-filled environments)	Once a year
Fuel filter or sediment bowl cleaning	Once a month (more frequently if gas tank is rusty)	Once a year
Fluid level check	Every use	Every use
Tire check	Every use	Every use
Tune-up	Each season	
Dressing	Once a year	
Spark plug gaps	Once a month	
Coolant change	Once a year	Once every other year
Hydraulic fluid change	Every year	Every other year
Transmission fluid change	Every year	Every other year
Brake and clutch	Once a year or as needed	Once a year
Linkage adjustment	As needed	
Engine accessory drive belts	Dressed and tightened twice a season	Dressed and tightened once a year
Repack wheel bearings	Once a year and at every tire change and as needed	Every three years
Inspect structural bolts	Every month	Once a season
Bearing preload adjustment	Every season	As needed

Some general notes:
- *If an operator's manual can be found, follow your manufacturer's recommendations.*
- *These recommendations assume covered storage. (Lubrication in particular should be done more often if the tractor is left out in the weather.)*
- *This assumes the tractor is operated in normal agricultural conditions. Particularly dry and dusty conditions require more frequent maintenance.*
- *These are minimum maintenance intervals. Performing any of them more often is encouraged.*

and other problems that can lead to trouble if not corrected.

To determine the best maintenance schedule, consult your tractor's owner's manual. A service or operator's manual will tell you how often to change the oil, transmission fluid, and hydraulic oil; when to lubricate certain components; and what style and type of fluids to use (this is particularly important for hydraulic systems). Many businesses specialize in providing replacement owner manuals if you've misplaced yours or never had one. Refer to the appendices for suppliers. The schedule on

Oil level dipsticks were non-existent on many antique tractors. This photo illustrates how you check the oil level on tractors without dipsticks. The higher of the two petcocks is the "full" petcock and the lower is the "add more" petcock. To check the oil, slowly loosen the lower petcock. If oil runs out, you are okay. If it doesn't, tighten it and loosen the upper petcock. Slowly add oil until oil begins to drip out of the upper petcock. Tighten the upper petcock when this happens, and you are done. You want to check these levels while the engine is hot. As with a car, you don't want to overfill the oil pan, so slowly add oil.

page 121 provides a general maintenance timetable.

LUBRICATION

There are only three important things to remember about tractor maintenance: lubrication, lubrication, lubrication (if I may shamelessly paraphrase an old saw from the real-estate industry).

While that greatly oversimplifies the matter, the majority of all maintenance procedures boil down to lubrication, cleaning up after it, or changing a fluid that lubricates. To apply grease to a lubrication point on a tractor, a one-way greasing valve called a "zerk" or "grease" fitting allows the grease to enter but not leave. For example, steering components have these types of fittings. Knowing where all these fittings are on your tractor is important, and applying grease is probably the first maintenance procedure you will perform. Here is where a copy of an owner's manual for your tractor will come in very handy. Most manuals will have a drawing of the tractor outlining every grease fitting and what type of grease to use if the fitting does not take ordinary grease. To apply grease, first clean the fitting completely. Then fit the tip of the hose of the grease gun over the grease fitting and pump the grease gun. The hose, if kept straight, will not come off while you pump grease but will be loose and easily removed when you are not pumping grease.

Often I am asked exactly how much grease should be applied to the assembly being greased. The dilemma is that too much grease is either wasteful or even potentially harmful, and too little grease won't lubricate the tractor. In most cases you are looking to pump enough grease to see a small amount escape from the joint you are lubricating. In some cases, such as high-precision assemblies, err on the side of too little. Too much grease in some high-precision assemblies creates high internal pressure during heat-up and can cause failure in the component. While this is rare in antique tractors, some devices such as universal joints should be sparingly but frequently greased. Regardless of what you are greasing, the most important step is removing excess grease from the zerk fitting and the joint immediately afterward. Grease traps dirt and moisture, and any excess left helps to introduce these contaminants to the component you are trying to protect!

FASTENERS AND FASTENING

Oddly enough, the proper way to tighten a bolt or screw is not entirely self-evident. Things like torque specifications, thread additives, parent material composition, and other things need to be taken into account. When reassembling parts, you should try to take the opportunity to clean out the threads of the fastener and the parent material, and replace the fastener if it is heavily rusted. If authenticity is important to you when replacing

Antique Tractor Tip

Fluid Battles
Water is your worst enemy when keeping up a tractor. Never store a tractor where it is exposed to the weather for an extended period of time. You should never store it outdoors.

fasteners, you will be pleased to know that many fastener companies (look in the Yellow Pages under Fasteners) have many of the square-head bolts and unusual clutch, pan, and airplane style fasteners used on antique tractors. For those unusual fasteners not stocked by these companies, there are some firms specializing in providing authentic hardware. They can be found in the appendix of this book.

If you are replacing a bolt, it is of utmost importance that you identify what type of bolt it is. If you arc pulling out a hardened steel bolt, such as a structural bolt or cylinder head stud, then you should replace it with hardened steel. In older tractors, hardened steel bolts are dark steel and will usually scratch a modern Grade 5 bolt. Otherwise assume it is nonhardened. Identifying the hardness of a modern bolt is simple: Count the radial slashes on the head of the bolt. Add two and that is the hardness. For example, a Grade 8 bolt will have six radial slashes. If there are no radial slashes, the hardness is either undetermined or it is a Grade 3 bolt. When replacing hardened steel bolts, ask a fastener supplier for Grade 8 bolts; otherwise, use Grade 5. When reassembling, it is best to clean out the thread of the parent material. If you are not replacing the fastener, you should clean off its threads also. Then lightly lubricate the fastener with a light oil, such as a 20- or 30-weight oil, and fasten to the torque specifications in the shop or owner's manual.

Here is a quick chart of uses for various fasteners:

Grade 8 fasteners: Cylinder heads, structural bolts (bolts that hold main structural pieces of the tractor together), any bolt or nut with a high torque specification (greater than 125 foot-pounds), axle nuts, rim-to-wheel bolts, and other bolts that you think may have structural implications. You can almost never go wrong using a hard bolt in place of a soft bolt. Use this grade if unsure.

Grade 5 fasteners: This is the most common grade of fastener used and you will find 90 percent plus of all bolts on your tractor can and should be this grade.

Less than Grade 5: For light-duty and decoration uses. Do not use this fastener in any situation where the bolt has to be relied upon to prevent injury or property damage.

Shear bolts: These bolts are designed to be very soft and break or shear when a heavy load is applied. This is done for safety reasons so equipment is not damaged nor people injured when shock load occurs. NEVER SUBSTITUTE REGULAR BOLTS FOR SHEAR BOLTS!

How tightly you should secure a bolt or screw depends entirely on the fasteners, the parent material, and the load the fastener will bear. The first thing to remember is that proper tightening requires a torque wrench. Bolts such as cylinder head bolts, bearing nuts, and the like should never be tightened without a

Every season, make it a habit to inspect the glass bowls found under the fuel tanks of nearly every antique tractor. Dirt and sediment leaves the fuel tank and is trapped in the glass bowl to prevent it from reaching the carburetor. After turning off the fuel flow with the small valve seen at the top left of the sediment trap, loosen the nut at the bottom and swing the wire sling out of the way. The bowl will stick a bit, but it should drop down with minimal effort.

torque wrench. The torque wrench will tell you, in the units of foot-pounds, how much force you are applying to the fastener. Comparing this to the manufacturer's guidelines will help you to apply the proper amount of preload to the bolt. The only other thing to remember about accurate tightening is that torque specifications assume the bolt will be lubricated with a light coat of oil. Any thread or bolt additive products will also serve as thread lubricant. I typically add 10 percent to the torque specs if I can't lubricate the bolt. If the parent materials are soft

Antique Tractor Tip

Grease Gun Hints
Always clean every grease fitting before each use. Wipe away excess grease after each use.

Alternators have one advantage to generators in that they charge at a constant rate, even at low engine rpms where most antique tractors operate. Here is a 12-volt alternator in place of the original 6-volt generator. When you make this upgrade you need to swap out your battery with a 12-volt battery and sometimes add a resistive coil to the primary circuit headed to the coil.

or structurally weak, then I use just enough force to snug things up. This is done by feel and takes just a little experience. You will probably strip a few threads and break a few parts overtightening before you get the hang of it.

Often when reassembling, circumstances require that extra steps be taken to ensure the fastener will stay put and not loosen over time. Or perhaps you want to try to prevent the fastener from rusting and galling into place, creating headaches the next time it

Antique Tractor Tip

Lube, Lube, Lube
When greasing the tractor, apply a small amount of engine oil to metal-to-metal contact surfaces such as hinges, unbushed shafts, and so on.

is removed. This is where fastener additives such as thread lubricants and fastener "glues" come into play. The most common brand-name for these products is Loc-Tite. These products range from simple lubricants to help assure bolt torque all the way to heavy glues that will require machine assistance to remove the bolt afterward. The most useful products in this class for antique tractors are thread sealers (that help prevent rust and galling), lubricants, and light glues that help stop problem fasteners from loosening from vibration.

Engine Maintenance

Here is where most folks focus their maintenance efforts, and wisely so. There is no system on an antique tractor more likely to suffer from the effects of delayed maintenance than the engine. All engine components are high-precision assemblies subject to extremes of temperature and stress. Excellent maintenance is critical to the longevity of your engine. Fortunately, engine maintenance is straightforward and not difficult. Certain procedures require a bit of time, but none are overly complicated. Following the advice in this chapter and the instructions found in the service and owner's manuals for your model of tractor will help you achieve the reliability and longevity that proper maintenance brings.

The most important thing you can do for your engine is to keep it properly filled with fresh, frequently changed oil. The oil is the lifeblood of your engine and any contaminants in the oil will

adversely affect your engine. You can't change oil too often, but you can change it too infrequently. If you change your oil frequently—for example, after less than 100 hours of operating time and less than three months between changes, you can change your filter every other oil change. If you change it less frequently, change your filter every time. When choosing an oil, be aware that detergent oils under specific circumstances can harm your antique tractor engine! This warning only applies to engines that have never had detergent oils used in them and have not been recently rebuilt. In these engines, sludge builds up, coating the inside surfaces of the engine. Over the years, this buildup can be extensive. If detergent oil is introduced all at once to an engine in this condition, there is a significant possibility this sludge will loosen and travel through the oil delivery system. This sludge may then block an oil passageway in the engine and part of the engine will become oil-starved. The component deprived of oil by the blockage will then fail catastrophically. Unfortunately, components that are served by the oil delivery system are critical precision assemblies that are expensive and time-consuming to replace. For example, main engine bearings can be ruined this way.

If you believe that your engine will be rebuilt in the near future, I would not bother using detergent oil until after the rebuild. If, however, your engine will continue to be used for the next several years, it is difficult to avoid using detergent oil, and if introduced to the engine properly, detergent oils can be added to your engine. Change the oil the first few times very frequently using nondetergent oils. Then change the oil the next time substituting one quart of detergent oil for one of the quarts of nondetergent oil. Continue using this for one or two more oil changes. Then change the oil using two quarts of detergent oil, and again the remainder would be nondetergent oil. Continue adding detergent oil to the oil changes, one quart at a time, until you are completely using detergent oil. Adding detergent oil to your engine in this manner will help to minimize any risk from loosening sludge. Another thing to realize is that detergent oil will clean all the sludge and grime buildup from behind seals, allowing them to start leaking. Many people have switched to detergent oils only to find that their antique tractor engine now leaks oil readily. Leaks from around the rear engine seal and valve guides are commonly revealed by detergent oil (the detergent oil isn't causing them, it simply is cleaning the grime that was masking the leak).

The first time you change the oil, I also recommend cleaning the pickup screen on the oil pump. These screens are usually nasty. You will have to remove the oil pan to do this. Adding oil to other parts of the engine will be necessary, too, and here is where a copy of the owner's manual will help. Some engines, for example, have oil cups on the starter and generator. Just a drop or two of oil every oil change is needed. Almost all antique tractors have a way by which you lubricate the distributor/magneto. Again, just

a drop or two is all it needs. Many tractors require oil to be added to the cooling fan shaft assemblies. Your owner's manual will clearly outline all the parts of your engine that need regular oil, and their location. The manual will also tell you what weight of oil to use in each instance. Consulting an owner's manual and using a little common sense will go a long way toward making sure every component on your tractor's engine is properly cared for.

COOLING SYSTEM

Radiator fluid should be changed once a year on regularly used tractors, and once every two years on infrequently used tractors. The fluid should be a 50/50 water/ethylene glycol mixture. This mixture will protect against freezing and boiling and will help to prevent further rusting and scaling of the cooling system. If your antique tractor has a water pump, no maintenance is necessary other than to inspect for leaks from time to time. Part of regular maintenance on cooling systems includes cleaning chaff and trash from the cooling fans of the radiator and straightening bent fins. Be sure the overflow tube is unobstructed. The only other maintenance concern for antique tractor cooling systems is to be sure the thermostat operates. Thermosiphon cooling systems may or may not have a thermostat. To check the operation of the thermostat, simply check for the presence of heat either at the top of the radiator or at the hose running to the top of the radiator. Heat should be present after the tractor has been worked for a while. The lack

Here are two common maintenance tasks on an antique tractor. The first is an oil change. The oil filter is located in the oil filter housing that is slightly top of center. Antique tractors did not have spin-on filters like today's engines and they had separate housings like these for the filter. The second task is removing the protective band from around the starter brushes (slightly below the center of the photo), inspecting the brushes, and blowing the dirt and dust from around the brushes.

of heat means the thermostat isn't functioning properly or was removed by a previous owner and needs to be replaced. Also, check your radiator cap if your antique tractor has a pressurized cooling system.

Antique Tractor Tip

Hot Oil Drain
Regardless of the type of fluid, try to drain fluids (engine oil, etc.) when they are warm. The old fluid will drain much more completely when it is warm.

Antique tractor exhaust designs are endless, but the most common is the straight-up exhaust. This fully galvanized muffler is clamped onto a pipe that is threaded into the exhaust manifold outlet. It should be long enough to put the outlet of the muffler well above the operator so he or she isn't breathing exhaust while operating the tractor. All straight-up mufflers, like this one, should be topped with a rain cap.

The cap should be able to hold the pressure in the system at or near the manufacturer's specification.

Auxiliary Drive Systems

The "driven" components on the engine, such as the governor, hydraulic pump, and the like are typically self-lubricated and need no maintenance along those lines. Older antique tractors do have external governors, and these may need to be lubricated. The integral governors usually pick up oil from the front cover, camshaft/valve area of the engine and require no

further lubrication. This is also true of engine-driven hydraulic pumps. Hydraulic pumps do not require periodic maintenance, but some light adjusting may need to be done to your governor. Adjusting governors can be done two different ways. The first involves adjusting linkage between the carburetor and the governor. This linkage increases engine speed under load and its length determines how quickly and how well the carburetor reacts to the governor. Unfortunately, there is no way to generalize how to adjust the governor to carburetor linkage. The parameters are different from engine to engine. Typically, though, the linkage is adjusted so the carburetor throttle plate is nearly completely open when the throttle lever is wide open. The linkage between the governor and the throttle lever is also adjusted with the throttle wide open, but this time the engine must be running. The throttle-governor linkage is adjusted so the engine, at top throttle, runs at a certain rpm. As I mentioned, the particulars for your tractor, such as the best places in the linkages to make this adjustment, must be gleaned from an operator's manual or a person experienced with your particular model of tractor.

Fuel and Air Delivery

The carburetor and fuel delivery system must be maintained. Fortunately, very little maintenance is needed. On gas engines, the maintenance includes replacing the fuel filter (most antique tractors did not come originally from the factory

with fuel filters, but this filter is commonly added over the years by the owners), adjusting idle speed, adjusting the fuel air mixture, and cleaning the air filter. Cleaning the air filter simply entails emptying the oil pan at the bottom of the filter, cleaning it thoroughly, and adding fresh oil. Adjusting idle speed requires a tachometer. Install the tachometer, idle your engine, and then turn the idle-speed screw until the tachometer reads the required rpm. Once that is done, you can adjust the idle-mixture screw. The idle-mixture screw adjusts how much fuel enters the air stream going into the engine. This in particular affects low-speed engine performance. Some carburetors do have high-speed mixture-adjusting screws. Their adjustment is simple and requires an owner's manual to find the proper initial setting and the proper adjusting procedure.

To adjust the idle-speed mixture screw, warm the engine up thoroughly. The idea is to find the setting that results in the fastest idle speed for the tractor. Very slowly turn the screw counterclockwise until the engine speed starts to slow down. The engine may speed up momentarily, but if you turn the screw long enough, it will slow down. Once the engine starts slowing down, stop turning the screw. Now start turning the screw clockwise until the engine speeds up slightly, and then starts to slow down again. The differences in rpm are very subtle and often I need the tachometer to tell me exactly what the engine is doing. Once the engine starts slowing down, stop turning the screw again.

More than likely you moved the screw approximately one-quarter to three-quarters of a turn to affect the engine speed. The engine ran its fastest somewhere between these two extremes. Slowly turn the screw back to this point and that is about where you should keep the mixture screw positioned. You may need to go back and make a further minor adjustment to bring the engine's idle speed to the maker's specification.

Diesel fuel delivery systems are difficult for the average antique tractor owner to service alone outside of a couple of small maintenance tasks. First, regular attention should be paid to the system's filters. There is at least one, usually two, filters that will separate and remove water and contaminants from the fuel. Since the diesel fuel pumps, metering apparatus, and injectors are high-precision components, make very sure that your fuel, fuel tanks, and filters are kept very clean. A repair shop specializing in diesel equipment can clean injectors for a very reasonable fee. Afterward, using an additive will keep them clean for years to come. Specifics on the types of filters and their replacement procedures can be found in the owner's manual.

EXHAUST

The exhaust system of antique tractors is simple and requires very little maintenance. The only maintenance task here is inspection. You are looking for exhaust leaks, primarily. Exhaust leaks can emit sparks, which can be very dangerous and must be prevented. This is especially true if you plan to use the

tractor around buildings, fuel, hay, or other combustible materials—which at some point you are sure to do. The common places exhaust leaks are found is at intersections of manifolds and exhaust pipe, the intersection of the muffler and exhaust pipe, and from around the exhaust manifold gasket. Occasionally, the heat, age, and rust eventually erode the exhaust manifold to the point where holes develop in the manifold itself. Look primarily at the corners and underside of the manifold to find these. Another effective way to find exhaust leaks is to run the tractor at full throttle in a very dark area. Then rapidly throttle down the tractor. Do this three or four times. Exhaust gas will occasionally flow or flame from a leak when you do this. If the leak is large enough, sparks will fly, making the leak quite easy to spot.

ELECTRICAL
Primary

If your antique tractor does not have any electrical component other than a magneto, you can safely skip this section. The primary electrical system is low voltage. For example, if your antique tractor runs on 6 volts, then all the electrical systems and components that run on 6 volts form the primary electrical system. The lighting, starter, battery, generator, and so on are all part of the primary electrical system. The portions of the electrical system that generate higher voltages for the formation of a spark at the spark plug are called the secondary system. Examples of secondary components are the coil, the high tension side of distributors, spark plugs, and so on. A diesel engine tractor, therefore,

Since antique tractors are used infrequently, you'll find it challenging to keep their batteries charged and ready to go. Just about any charger will do, but I recommend you get one like this unit that has an automatic mode.

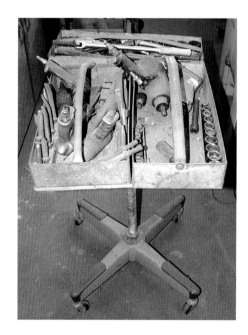

A tool cart is a very handy thing to have. Here a cart is used as a place to store all the tools being actively used on a project. The great thing about these is there is no need to store the tools every time you stop for the day. You can keep them beside the project in the cart until the next time you work on it. These are great for storing parts during the project.

does not have a secondary system. The primary electrical system requires little maintenance, and most of that revolves around caring for the tractor's battery. The battery must be topped off with distilled

Adjusting valves is a common maintenance procedure that should be done every few years, or more often if your antique tractor sees any field work. While your tractor manual will list the specifics for the model, such as valve clearance specifications, this series of photos will give you a great start in understanding the procedure. This engine has quite an unusual valve configuration because the valves are side mounted. Steve Sewell starts by carefully prying off the valve covers. Be sure not to damage the seal, so it can be re-used. The cork seal under the cover is in good shape and can be re-used. Note the cover has stand-offs to prevent the bolts from collapsing the cover when reinstalling, but most engines don't have these. Be careful not to overtorque the valve cover bolts when installing valve covers; you may possibly dent or deform the covers. Steve begins the adjustment by rotating the engine by hand (loosening the spark plugs so it is easier to turn is a good idea) until the cylinder associated with the valve he is about to work on is at top-dead-center on the compression stroke. First loosen the locknut with a wrench while holding the adjusting screw in place with a screwdriver. He next places a feeler gauge of a thickness that matches the clearance (known also as "lash") specification and turns the adjusting screw until a slight effort is required to retract the feeler gauge. Using the screwdriver to keep the adjusting screw perfectly still, he tightens the locknut. After tightening, check the clearance again to make sure tightening the locknut didn't turn the adjusting screw and alter the clearance.

water on a regular basis and should be kept charged up during periods of nonuse. The connections should be kept clean and any scaling or corrosion should be neutralized with baking soda when needed. Be sure not to get any of the baking soda solution in the cells of the battery. The other areas of maintenance for the primary electrical system are keeping connections clean and tight and inspecting the condition of the wires from time to time. Another maintenance procedure that should be performed is a voltage check of the charging system. The voltage of the primary ignition system, when measured across the battery posts at full throttle, should be about 5 to 20 percent higher than the nominal voltage of the system. For example, if your antique tractor is a 6.0-volt system, then your voltmeter should read about 7.0 volts during this test; if your tractor is a 12.0-volt system, then the voltage will read about 13.3 to 14.0 volts.

Secondary

The secondary electrical system is that part of the electrical system responsible for creating a spark at the spark plug. This part of the electrical system, also called the ignition system, requires much more maintenance than the primary electrical system. The overall idea behind the ignition system is to create a very high-voltage current and send it to the spark plug at the right time. I think you will find that 75 percent of all engine problems relate back to poorly performing ignition components. Maintaining the system is one of the trickier aspects of antique tractor maintenance, but these procedures are more important than most and we will try to cover them as completely as possible here.

MAGNETOS

Antique tractor owners have a love-hate relationship with magnetos. Magnetos are one of the distinguishing characteristics of antique tractors and speak directly to those emotions, fascinations, and motivations that drive us to collect and use antique tractors. Most of us find them fascinating marvels of engineering from bygone eras. Magnetos also can be notoriously difficult to fix, find parts for, and understand. They often do not fail completely, but will only fail intermittently, making them difficult to troubleshoot. Fortunately, when they work they work great. Maintaining them is easy too, and this section should help you do just that. There are several maintenance procedures involved with magnetos that mirror exactly the procedures performed on battery coil ignition systems. These procedures are gapping the points (setting the dwell angle), maintaining spark plug and coil wires, timing the engine, and so on. When you are ready to perform these procedures, skip ahead to the battery ignition section to learn how.

The tasks specifically associated with magnetos are cleaning, lubricating, checking the general operation of the impulse coupling, and verifying that the magneto is capable of producing a strong, healthy spark. Cleaning the magneto is straightforward, and requires little more than common sense. Most manufacturers recommend a light cleaning every few hundred hours and a thorough cleaning every 400 to 600 hours. The thorough cleaning should be down off the tractor at a bench and should include the drive and impulse coupling area of the magneto. The preferred solvent for cleaning the magneto is a spray form of electrical cleaner. This substance will cut grease, but will be friendly to your magneto. Some of these solvents will dissolve the lacquer finish on the coil, so test first! If a special electrical cleaner is not available, then kerosene will work well; just be careful to keep it restricted to the mechanical areas of the magneto.

To lubricate the magneto, simply place a drop or two of oil in the grease/oil cups(s) of your magneto about every 50 hours. This is one component where you must be careful not to overlubricate! There should be no need to oil parts of the magneto other than the cups. To check the impulse coupling, move the tractor to a quiet spot away from noise and distraction. Have a friend hand-crank the tractor with the kill or ignition switch off. While the engine is turning, you should hear noticeable clicks from the magneto. These clicks should sound sharp, strong, and metallic. If the clicks sound soft, slow, spongy, or intermittent, then the impulse coupling is dirty and will need to be cleaned. While cranking the engine, also check the condition of the spark of the magneto. To do this, remove the coil wire from the magneto at the distributor. Situate the wire so the end of it is almost, but not quite, touching a good ground—maybe the base of the magneto, an exposed area of the engine block, or the like (being careful not to place your fingers on or near the leads). Then slowly turn the engine over with the ignition on. Watch the spark as it jumps between the end of the coil wire and

Some antique tractors use the engine block as a structural part of the tractor. In other words, the front axle or pedestal bolts to the front of the engine block and the back of the tractor bolts to the back of the engine. The block is part of the frame. Here you see something different. This is known as a bathtub frame, and the engine (missing here) sits in this frame. In this case, removing the engine is quite a bit easier.

the ground. The spark should be bright and the color should be blue. If the color looks yellowish or orange, and seems dim and weak, then your magneto may need attention. In all likelihood in this case, either the magnets are weak or the coil is on its last legs. Check the spark several times with different grounds to verify your suspicion that it is weak.

DISTRIBUTOR AND BATTERY COIL

The maintenance for a distributor and the distributor portion of a magneto is multifaceted and involves many procedures, parts, and tools. From here on out, when I refer to a distributor, I am including the typical distributor and the distributor portion of a magneto. The most important thing for a distributor ignition is

In the valve lash adjustment photos we saw a slightly different valvetrain assembly for a flathead engine—it had tappet arms for actuating the valves. In this engine, we see the more traditional approach taken by engine and tractor manufacturers. This engine has small barrel-shaped tappets that ride up and down on the camshaft lobes. At the tops of the tappets are adjustable strikers that move the valves up and down.

When working on antique tractors, moving, setting up, and being able to access large parts and assemblies is important. Here a forklift is being used as a hoist for an engine while it is being sandblasted. While most of us don't have access to a forklift, solving the problem of how you will move and raise heavy parts, such as front axles, wheel weights, or wheels, is something you need to consider, and ingenuity such as this will be needed at times. For this particular circumstance, most of us use engine hoists, but be forewarned: they often don't lift high enough.

to keep the components tuned up. By a full tune-up I am referring to replacing spark plugs, spark plug wires, secondary coil wire, distributor cap, rotor, points, condenser, coil, and—if your tractor has an external resistor— the resistor as well. Most of the time, folks will closely inspect and test various components before replacing them. For example, distributor caps very rarely go bad and the cap is only replaced when the contacts inside it look burnt and rounded. If, however, your new tractor hasn't seen maintenance in quite a while, or perhaps you have ignored one of your older tractors, spending a few extra dollars to perform a complete tune-up may be money well spent. I perform a tune-up every year, trying to keep components like the coil, cap, rotor tower, and the like if they seem to be in good condition. Every other year or so I will perform a complete tune-up.

There are certain procedures you must follow to do a tune-up. The first will be inspecting, cleaning, and gapping the spark plugs. To clean the plug, simply use a brake- or carburetor-cleaning solution. Unless the plug was very clean to begin with, it will not come completely clean, but should be noticeably cleaner. There are sandblasting devices on the market to clean the plug, but I recommend not using them. With these devices, it is much too likely that a grain or two of sand will lodge between the insulator and the case of the plug, only to drop after the plug has been reinstalled and the engine begins running. I can't

believe that grains of sand inside your engine would be ideal, so I prefer to clean plugs with a liquid solution. To regap the plugs, you must first lightly etch the electrode and gap arm with a points file (a small, fine file used to file ignition points), or as a second choice, a very fine grade of sandpaper. I prefer to use wet or dry sandpaper, as the particles remain on the paper and don't seem to fall off onto the plug. Then you can gap the plugs using a gap tool. Open or close the gaps as needed by bending the electrode to create a gap equal to the recommendation found in your tractor's owner's manual.

The second procedure is setting the ignition point gap. The points are found inside the distributor, underneath the distributor cap. (Often there is another cover under the distributor cap you must remove to reach the points.) The points open and close as necessary, creating and breaking the magnetic field in the coil that ultimately creates the spark. How long the points stay closed is important, and setting the gap ensures that the current that is building the electromagnetic field stays on for the proper length of time. To set the gap, lightly file the point with a points file or, as a second choice, lightly sand with a fine grade of wet or dry sandpaper or emery cloth. Be sure to remove any dust or grit, then begin setting the gap. To do this, rotate the engine by hand (99 percent of all tractor engines are designed to rotate clockwise when looked at from the front; check with your owner's manual). Continue turning until the points are open to their widest. This will be when the cam follower of the points sits on top of one of the high cam lobes of the distributor shaft. Measure this distance using a feeler gauge. Compare this with the recommendation in the owner's manual.

To change the gap setting, loosen the lock screw and turn the adjusting screw until the proper distance is achieved. The proper gap should create very light friction on your feeler gauge when pulling the feeler gauge out, but the gauge should still very easily pull out. While holding the adjusting screw still, tighten the locking screw. Double check the gap and readjust if necessary. One other maintenance procedure on the points should be considered. Often the proper gap is expressed in terms of "dwell angle." This is the amount of distance, expressed in the number of degrees around the distributor shaft, that the points remain open. This test ensures your gap is correct, your condenser is healthy, and the cam follower on the points is not overly worn. In short, the dwell angle test verifies the results of your maintenance procedures and should be performed if you have a device to do so. If not, and if after your procedures the tractor runs fine, the test may be safely skipped.

The next procedure to follow when tuning up the ignition system is to time the engine. By this, we mean we are going to make sure the spark at the spark plug occurs at the proper instant during the engine cycle. While intuitively it seems that the spark should happen right at the instance when the piston is at the top of the compression stroke, the fact of the matter is that the spark must be initiated slightly before then. Therefore, all engine timing is expressed in the number of degrees before top dead center, BTDC. This is usually about 2 to 8 degrees, but it can be slightly more than that. Once again, your owner's manual will have the proper timing specifications. First, a couple of other tasks must be performed before you time the engine. You must first verify that the engine is idling at the proper rpm and the fuel-to-air mixture is set properly. That was discussed a little earlier in the chapter. You must also verify that the timing marks on the engine (typically on the flywheel or on the crankshaft pulley) exist and are visible. In addition, you must find the "index" mark or pin. This is the reference point with which the timing mark will line up when the engine is properly in time. The timing mark is almost always stamped into the pulley or flywheel, and it will show up best during our timing procedures if you fill the recesses of the timing mark with white paint. I also like to use chalk. Simply rub the chalk back and forth over the timing mark, then lightly wipe the mark with your finger. This will remove the excess from around the mark but will leave the chalk in the mark. Sometimes it is helpful to paint or mark the reference point as well.

Then run the tractor until it is at operating temperature. Shut down the tractor and install a timing light to the spark plug and spark plug wire of the number 1 cylinder. Inductive timing lights are best, but most require a 12-volt battery—

Many antique tractor radiators are three-piece systems bolted together and not a solid assembly where the tanks are soldered to the cores like modern vehicles. As you can tell from this John Deere radiator, you have a heavy cast-iron top tank, a thick radiator core, and a cast-iron bottom tank. These are bolted together with gasket material at the junctions. Leaks often occur at these seams, so if your radiator is leaking, be sure you check the seams before you despair and assume you will have to get a costly radiator recore job done.

something most antique tractors don't have. Noninductive lights are not as bright, so performing this procedure in the shade or in a dark shop is important. Install the light, start the tractor, and shine the light at your reference point. If the timing of the tractor is at least close to the proper specifications, you will see the timing mark show up in your light. Since the timing light in essence is a strobe light that flickers on and off each time the engine initiates a spark, the timing mark will seem still and constant. If the engine is in proper time, then not only will the mark show up, it will also line up with the reference.

If the mark doesn't line up with the reference point, then you need to loosen the locking bolt on the distributor or magneto and very slowly rotate the distributor. This all must be done with the engine running, so having a helper rotate the distributor is necessary for safety. While the proper direction to turn the distributor is something you can ascertain by studying the rotation of the distributor shaft, it is much easier to just pick a direction and start turning. If the direction your helper is turning the distributor is the correct direction, then your timing mark will start moving toward the reference point. If the rotation direction is wrong, the mark wanders away from the reference point and your helper must change the direction of rotation.

If the mark doesn't show up, but the engine seems to be running reasonably normal, then either the mark, the reference point, or the light is at fault. More than likely, the mark isn't visible enough under your current conditions. Move the tractor to a darker area, repaint the timing mark, and try again. If the mark still isn't showing up, then either the pulley is incorrect for the engine, or if your timing marks are on the flywheel, the flywheel was bolted to the engine incorrectly (on some engines it is possible to bolt up the flywheel incorrectly, while it is impossible to assemble the proper pulley incorrectly). The last possibility is that your reference point is incorrect. Be sure your reference point is indeed the proper point. If those three things check out, check your light and make sure it works properly on another engine.

It may be firing sporadically, making reading the mark impossible. For the mark to not show up, one of these four things must be the culprit.

VALVETRAIN

The valvetrain requires periodic maintenance. About once a year the "lash," or play, in the valvetrain must be adjusted. How you adjust the lash depends entirely on the engine you are performing the adjustment on, and a copy of the owner's manual is a must. The manual will tell you two important things. The first is the clearance, the second is whether this procedure needs to be done with the engine hot or cold. There doesn't seem to be any rhyme or reason among manufacturers, but most require that the engine be brought up to operating temperature first before adjusting the valve lash. Before you start, be sure to inspect the operation of the valves by rotating the engine by hand. You are hoping to see valves that move up and down freely and smoothly, and you are hoping to hear no sounds and see no extra sludge, grime, or dirt in the valvetrain area. To begin, bring the tractor up to operating temperature if that is what your manual calls for, then shut down the engine.

If your engine has a valve-in-head design, then remove the valve cover from the top of the engine. If your tractor has a flat head engine (also known as an L head), then you must adjust the valves through a valve chamber cover in the side of the block. Once this is removed, turn the engine until the number 1 cylinder is at top dead center. At

this point, both valves for cylinder number 1 should be fully closed, and there should be clearance between the tip of the valve stem and the mechanism that pushes the valve open. In a flat engine, the mechanism will be a barrel-shaped object called a tappet or cam follower. In the overhead-valve engines, the mechanism will be a rocker arm. The clearances for the intake valve and the exhaust valve will be different, so be sure to identify the valves before adjusting. To determine this, simply follow the exhaust and intake manifold ports.

Once you have the right valves identified, you adjust the clearance by adjusting the striker of the tappet or rocker arm. These invariably are arranged so that the striker screws in or out, adjusting the clearance. There is a nut on the striker that locks it into place. To adjust the clearance, place a feeler gauge between the striker and the tip of the valve stem. Then loosen the locking nut and turn the striker in or out as needed to adjust the clearance. Then, while holding the striker still with the wrench or screwdriver, tighten the locking nut. If you tighten the lock nut without holding the striker screw still, the striker will turn as you tighten the locking nut, putting the valves back out of adjustment. Having a helper is handy when adjusting valves. This step is repeated for every cylinder.

Chassis and Drivetrain

The nonengine components of antique tractors are usually quite simple to maintain. Like the engine, they require periodic lubrication, but unlike the engine, there is very little in the way of adjustments or other complicated maintenance. The systems you need to pay particular attention to are the hydraulics, transmission, PTO/belt pulley, and wheel/axle bearings. There are many other items to maintain, but these are the critical systems.

CHASSIS LUBRICATION POINTS

Bearings and bushings throughout the tractor require periodic maintenance that typically means cleaning bearings and lubricating them. Bushings typically are not removed, and if they aren't removed, they aren't cleaned other than having their external seams wiped off. Some bearings are removed and cleaned, however, and then lubricated, or "packed," with grease and reinstalled. The bearings in the front axle, steering components, and implements and trailer wheels are usually of this type. The bearings in the transmission and rear axles and final drives are of the "oil bath" type. That means they are not removed and stay continuously lubricated through a fluid within the housing they are in. For example, the bearings of rear axles are lubricated by the fluid in the differential or the final drive houses. The fluids also keep the bearings clean by washing the contaminants away.

To clean the bearings, you must first remove them. Some bearings are exact fit. That means the bearings fit onto the axles or spindles exactly, and the bearing

can be removed by hand or with only gentle persuasion. Some bearings are press-fitted. That means the bearings are placed onto the parent stock with force, and force is required to remove them. There are a number of bearing pullers available at tool stores and tool rental yards that should do the trick. It has been my experience that the bearings requiring regular maintenance are exact fit; the bearings not needing regular maintenance are press-fitted. There have been exceptions and you should not buy or rent a tool until you have checked your bearings.

Thoroughly wash the bearings in parts cleaners. There are many good parts-cleaning solutions at your auto parts store. You must make sure to remove all traces of old grease and dirt. Once the bearing is clean enough to eat off, then you will need to repack it. There are several gadgets at your auto parts store for packing bearings, but I prefer packing them by hand. To do this, simply get a golf-ball-size glob of grease in the hollow of your hand. Then slowly scrape away at the pile of grease with the bearing, using the palm of your hand to push the grease into the recesses of the bearing. Turn the bearing and repeat. Do this completely until no air pockets exist in the bearing. It is important to get grease in every nook and cranny of the bearing. The exterior of the bearing should not be overpacked with grease, however, and just a light 1/8-inch film on the outside is sufficient. To make sure you haven't overpacked it, turn the bearing slowly a time or two. Excess grease will bulge

Antique Tractor Tip

Dust Devils
Cleanliness is important when maintaining several systems on tractors. Primarily, engine and hydraulic systems need to be repaired and maintained in locations where dust and dirt are controlled.

out. Now lightly wipe the grease on the outside of the bearing and reassemble. The service manual for your tractor should be able to illustrate which bearings need packing and which do not.

Other bearings, such as front axle thrust bearings, may come from the factory permanently lubricated and therefore not need lubrication. These will be omitted from the service information in the manuals. Bushings typically receive their grease through a zerk fitting. Chains, either drive chains or power transmission chains used to couple shafts, usually require frequent light oiling, but some are greased; whether you should oil or grease them will be outlined in the owner's manual. When in doubt, use frequent light oiling. The last areas to be lubricated are the metal-to-metal contact areas. These are components that rub or mate together, yet do not use bushings or bearings to control wear or positioning. Occasionally they have zerk fittings for the application of grease, but often just a light coat of oil occasionally applied is all they need. Typical

Small, simple improvements can save much aggravation and time. Here a small rubber slip is suspended in the hitch frame. This prevents stems from hay being twisted up into the PTO shaft of the mower, which creates a mess that prevents the PTO from sliding in and out as you make turns.

examples would be the pivot points for levers, throttle linkage joints, and steering column braces.

POWER TRANSMISSION CLUTCH

The clutch requires periodic service to ensure longevity and to prevent costly repairs. Clutches are time consuming to remove and replace in most cases, so this is an area you will want to make sure gets attention during maintenance. The two most important things to make sure of is clutch pedal free play and lubrication of the clutch fork shafts. The driveshaft may also ride on a pilot bearing or bushing found near the clutch, and this would be the time to maintain that bearing as well. Your owner's manuals will outline exactly where and how the clutch shafts and linkages need to be lubricated. You will find one of three scenarios for most tractors. The remaining part of this section outlines them.

The first is the external clutch packs of the Waterloo two-cylinder John Deeres. These require little maintenance but need adjustment of the clutch lever at the clutch pack itself. Adjusting the clutch involves turning, synchronously, nuts that take up the slack between the clutch components caused by wear. When you are adjusting the clutch, you should feel, at the lever, the clutch engaging after a small distance of free play. The clutch should engage smoothly and surely and there should be no jerkiness or lack of positive engagement. When disengaging the clutch, you should feel a definite snap as the clutch pack disengages.

The other common clutch linkage design has two variations. Both use a foot pedal that activates a clutch fork. The shaft the fork rotates on requires periodic lubrication. Some tractors, such as Farmall, have a clutch access hole in the bell housing portion of the torque tube that allows access to a grease zerk fitting that lubricates the shaft and throw-out bearing. The second style involves a shaft that protrudes from each side of the torque tube, and there is one grease fitting on one side, or occasionally there are two fittings, one on both sides of the shaft. Adjusting these clutches usually involves changing the dimension of a connecting rod between the clutch pedal and the rockshaft that actuates the clutch. This rod will have threads at one or both ends. Turning either the connecting rod or the nuts on the threaded portions will move the rod and change its dimension between the pedal and rockshaft. Your owner's manual will give you specifics as to how much free play there should be in the pedal, but the most important thing is that there is at least some free play. Most tractor manufacturers called for 1 to 2 inches of clutch pedal free play for their tractors.

The clutch fork rides on a thrust bearing, which in turn rides on the clutch fingers that disengage the clutch. These thrust bearings are typically permanently lubricated and do not need attention. While most tractors have a pilot bushing that needs no maintenance, some tractors have pilot bearings. These usually require lubrication. Pilot bearings maintain the alignment and stability of the drive shaft and are usually found somewhere near the clutch. If your tractor has pilot bearings that need regular attention, take care of them when lubricating the clutch components. The owner's manual will tell you the specifics you need to know for your tractor.

TRANSMISSION

With the exception of inspection and some other minor procedures, such as cleaning out transmission breathers, the vast majority of antique tractor transmissions never need any maintenance attention other than changing the transmission fluid. Deciding what type of fluid you should use, however, could be a complicated process. Some transmissions, such as those of the Waterloo two-cylinder

John Deeres, simply share oil with the engine. When you change the oil for the engine, you are changing oil for the transmission. Other tractors, such as Massey-Harris, Farmall, and Allis-Chalmers, call for typical 80- or 90-weight gear oil. Some others, such as the Ford 8N, Massey-Fergusons, and later Farmalls, use the transmission fluid as the reservoir for the hydraulic system, so hydraulic oil is the oil required in the transmission. In particular, these types of transmission must NEVER have detergent-based fluid used in them. Unfortunately, many of the fluids recommended by owner's manuals are no longer being made, or are no longer being made under the same trade names. To decide on the right fluid, first consult your owner's manual. If the fluid called for is easily recognizable in its modern form, then simply follow the manual's suggestion. If the manual calls for 30-weight engine oil, then simply use modern 30-weight nondetergent oil (in the transmission, multiweight engine oils are not recommended). If the owner's manual calls for 80-weight gear oil, then simply use the modern equivalent (80w-90-weight multiviscosity oils are okay to use in antique tractor transmissions). If the owner's manual calls for hydraulic fluid, then the appropriate type of tractor hydraulic fluid can be used. This fluid is found at most farm supply stores. Here is where neighbors and the various local supply stores can be of assistance. They will know of the local brands of fluids that folks are using for their tractors.

There are a few other considerations when changing the fluid of the transmission. The differential, PTO, belt pulley, and final drive housings may or may not share fluid with the transmission. If they don't, you must be sure to change their fluids whenever you change the fluids of the transmission. Like the transmission, these components usually don't require any other regular maintenance, but you should inspect them at this time too. The belt pulley, especially if it is the paper type, should be closely inspected for rot and tears. All belt pulleys and PTO shafts should be grasped with both hands and shaken vigorously perpendicular to the axis of the shafts. If the pulley moves back and forth very slightly, or not at all, all is well. If it wobbles in its bearings, the bearings are excessively worn and repairs should be scheduled at the earliest convenience. Other inspections would include looking for excessive leaks around seals or around the power transmission housings, cracks in the housings, and loose mounting nuts and bolts.

WHEELS AND TIRES

Rims, wheels, and tires require periodic maintenance, and since rear tractor tires are one of the most expensive parts on an antique tractor (minimum $300 to replace a set of two), proper maintenance just makes good sense. The tires should be cleaned thoroughly once a year and a rubber preservative applied. The tire pressure should be checked frequently. Often many antique tractors have fluid placed in the tube of the rear tires instead of air. An ethylene glycol (antifreeze)

Portable air tanks save huge chunks of time if you plan on using your antique tractor in the field. With this in your truck, you'll always have air to inflate the tires of your tractor or implements when you aren't near your shop.

and water mix is added nowadays, but in the past a salt water solution of calcium-chloride and water was used. This solution, if it ever leaks out of the tube, will rust the rims of your tractor beyond repair, so many people drain this solution and add antifreeze instead. The rims and wheels of tractors tend to get a lot of abuse, and it pays to clean them thoroughly once a year and touch them up with paint. Of particular importance is making sure that

Antique Tractor Tip

Foam Tire Filler
If you operate your tractor in areas where its front tires are prone to puncture due to thorns, consider filling the front tires with a foam product that is sprayed into the tubes of the front tires. These foam products harden after curing. Most tire companies can help you find and install this product.

This is a typical pressure plate/friction disc clutch assembly. The disc in the front has a splined hub that rides on the splined shaft that leads into the transmission. The plate in the back is the pressure plate and is attached to the flywheel of the engine. Please note the two visible fingers found in the opening in the center of the pressure plate. These fingers are showing excessive wear and a new pressure plate or repair to the old fingers is needed.

the area of the wheels behind the wheel weights (if it is equipped with them), stays rust-free. Water tends to congregate behind the weights and cause rust.

HYDRAULICS

Hydraulic components are precision parts, and cleanliness is next to godliness when dealing with them. How you maintain this cleanliness,

Antique Tractor Tip

Implements Need Love, Too
Don't forget to frequently service your implements.

and how you change the fluids of the system, are dependent on the make of the tractor. Hydraulic systems vary considerably between manufacturers, and even between models of the same make. Making sense of your hydraulic system simply means understanding a little about how it works. Hydraulics are really fairly simple on a theoretical level. They consist of a reservoir, pump, piping, control valves, and a "ram." The reservoir holds extra hydraulic fluid to accommodate the fluctuation in amount of fluid actually in the system at any given time. The pump provides the push the hydraulic system needs to raise and lower implements. Piping carries the hydraulic fluid back and forth between components. The control valves control where the fluid goes, and a ram is the device that converts hydraulic work into mechanical work. Along every step of the way in this system, and in every component, dirt and grit can cause massive failures. This is especially true of the newer high-pressure hydraulic systems. You should change fluid in the reservoir on a regular basis, being very careful not to introduce contaminants in the process. Other maintenance procedures include keeping dirt and grime off the hydraulic piston rod, control valves, and, if your tractor is equipped with remote hydraulics, the remote ports. There is no adjustment with most hydraulic systems.

Crawler Issues

There are some additional maintenance procedures performed on crawlers that are not

performed on wheel tractors. The undercarriage, which consists of the rollers, sprockets, tracks, and other components, all require lubrication from time to time. In fact, the importance of maintaining the undercarriage is a close second to the importance of maintaining the engine. Undercarriage systems that are worn too far are impossible to fix without replacing a lot of expensive parts. Preventing that wear will save you a great deal of time and money in the long run. To make these components last requires lubricating them every day they are used. Like wheeled tractors, your greatest ally in maintaining a crawler will be the owner's manual, since that is the only book that will completely cover all the items you need to address.

Besides the rolling components of the undercarriages, there are two other related systems to maintain. The steering systems differ from manufacturer to manufacturer, but they fall into two categories: clutch/brake steering and differential steering. While the differential steering system is lubricated as part of the transmission and does not require additional adjustment, the steering levers and linkages require inspection, lubrication, and occasional adjustment. The clutch/brake system of steering requires lubrication and occasional adjustment. The track tensioning mechanism must be kept clean and well repaired. One of the most aggravating experiences with a crawler is to have a track slip the sprockets, and maintaining the tensioning system is important to prevent this.

Troubleshooting

One of the things I enjoy most about working with old equipment is its simplicity. Antique tractor engines take the basic concept of internal combustion and put it into practice in very direct, intuitive ways. The only problem with these engines is that they were built in a time when this technology was new to civilization, and many components, especially carburetors and magnetos, could be unreliable, prone to constant maintenance, and inefficient when they did work. Another problem is that these tractors are in the twilight of their years, and until the engines receive the complete rebuilds they usually need and deserve, the engines require more maintenance and general tinkering than they did when they were new. Of course, many owners consider it part of the fun to tinker with them and keep them running well. While these engines will sometimes confound you during the process of troubleshooting and repair, hearing your engine come back to life after a particularly difficult problem brings a tremendous sense of accomplishment.

Deciding what you should tackle yourself and what you should let someone else handle is a bit of a dilemma. Performing your own

The teeth along the edge of the flywheel, such as the ones shown here, are the teeth the starter of the engine engages. These teeth wear over thousands of starts. While they are showing some rounding and wear, they are in good enough shape to continue using with confidence, especially if this tractor will never be used for regular work.

Antique Tractor Tip

Rule of Thumb #1
If your tractor runs poorly, the problem is most likely a fuel, air, or compression concern. If it won't start or stay running, look to your ignition for the troubles.

The Art of Troubleshooting

Before you dive in, I would like to share with you some of my methods for successful troubleshooting. I have developed a number of insights and techniques over the years that I hope you will find helpful in your development as an antique tractor mechanic. Interestingly, the entire list has to do with approach and attitude, not particular skills or quantity or variety of tools:

▶ Believe you will eventually fix it.

▶ Commit to fixing it yourself.

▶ Do everything you can to understand how the system is supposed to work.

▶ Logically subdivide the system causing the problem into basic functions and components that can be tested individually.

▶ Use the process of elimination to dismiss those components that by definition cannot be causing the problem.

▶ Use reverse transversal (start at the problem and follow the sequence of events and inputs/outputs in reverse) to determine which of the components may be suspect.

▶ Decide in what order you want to test all the remaining suspect parts and systems. (The order of testing may vary; the important thing is to decide

repairs, especially many minor ones, is within every owner's grasp. I do recommend that everyone try to perform his or her own repairs if he or she feels the least bit inclined to do so. There are times when it makes sense to hire out the repair, though. For instance, to perform a repair you haven't done before, you will need time and resources to learn and work through the problem. If you are pushed by time or weather, investing in a learning experience is a luxury you probably can't afford. When you hire out the repair, you will find that professional mechanics usually can do the work much faster, or if the tractor is stranded in a remote field, a mobile mechanic will be able to carry out the repair better than you will. Many owners who hire out the work don't have the space or prefer not to invest in the tools. Whatever the reason, there are instances that hiring out the work makes sense.

Just because a repair is difficult doesn't mean you shouldn't consider trying it yourself. The difficulty of the repair should only be a determining factor if you are completely unsure of your mechanical abilities or have no experience with mechanical repairs of any kind. Books such as this, courses at community colleges,

and helpful friends and neighbors will be able to give you the intellectual tools you need to make the repair in most cases. Very little in antique tractors is conceptually difficult. Of course, you will have to learn as much as you can, and even after the learning process, you may still feel too uncomfortable to try the repair, but don't dismiss a repair on the grounds of apparent difficulty until you have looked completely into it. Chances are, if you have decent manual dexterity, a bit of patience, good work habits, and some sense of adventure, you could complete most of the repairs your tractor may require.

If your tractor is acting up and you decide to try the repair yourself, then take a deep breath, read through this chapter, and be safe. You will be surprised how rewarding it is to do your own repairs! Just remember, safety is your responsibility. If at any time during the repair you start questioning the safety of a situation, back away from it! Consult the safety and shop procedures chapters of this book, check your owner's manual, or talk to an experienced mechanic until you are satisfied you have made the situation safe. Never work on something where you are unsure of your safety.

on a game plan up front when logic isn't narrowing the possibilities far enough.)

▶ Perform these tests one at a time and remain organized throughout the entire process.

▶ Realize that our abilities to understand and comprehend these systems are sometimes inadequate and imperfect. For that reason, be open and receptive to hunches and instincts. In fact, instincts have a tremendous role.

▶ When all else fails, seek help from friends, relatives, and other books. The answer is yours to grasp, and finding it will only make you that much more effective the next time you need to troubleshoot.

Using the above advice, while it will occasionally result in the definite identification of the cause, will more often simply narrow the possible causes of a problem down to two or more systems or parts. How you discern between them and test them is up to you. Ideally, the first part to be tested should be the component most likely to be causing the problem; then test the part with the second highest probability of causing the problem, and so on. There can be, and often are, other criteria for determining the order of tests. Sometimes it is financial, because you lack the proper test equipment or maybe you have in your shop known good replacement parts that can be used for testing. In short, do not allow yourself to be

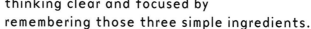

Antique Tractor Tip

Rule of Thumb #2
There are only three things needed for an engine to work well: fuel, spark at the right time, and compression. When troubleshooting, remember to keep your thinking clear and focused by remembering those three simple ingredients.

flustered by the possibility that the component you suspect is expensive or very difficult to test. Simply alter the order of testing, and leave that component until last. This process has the benefit of ensuring that everything but the last component is functioning properly.

DIAGNOSTIC TOOLS

When troubleshooting a tractor, a few diagnostic tools are handy to have that are not used in any other capacity. The first is a multimeter. This electronic device is very handy to have to trace electrical problems. Other tools for troubleshooting are:

• Compression gauge
• Radiator pressure gauge
• Hydraulic pressure gauge
• Mechanic's stethoscope

All of these tools can be rented, and certainly don't need to be bought until they are needed. There are many other tools that are indispensable to the troubleshooting process, but they are tools that you would own and use for purposes outside of testing and

troubleshooting. An example would be a battery hygrometer. This device will accurately report how well the battery is charged. While handy for troubleshooting a tractor that isn't charging the battery, it is also used regularly while maintaining the tractor to ensure that the battery is in good health.

Complaints by Systems

WHAT AN ENGINE NEEDS TO START

For an engine to start, it must be able to generate good compression, have sufficient fuel, have a spark in the cylinders at the right time, and have a healthy starter and battery to spin it up (or a healthy strong operator in the case of a manual start tractor!). Here are a couple of pointers to keep in mind when troubleshooting an engine that won't start or remain running:

First, refer to the table on page 140 to determine which system requires testing. The table is not all-inclusive, and very strange aberrant behaviors can leave one unsure of how to proceed. If your starter is not spinning up the engine adequately,

COMPLAINT	PRIMARY SYSTEM TO TEST
Engine does not spin up	Starting system
Engine does not try to fire	Ignition system
Engine tries to fire, but does not successfully start, or starts but does not remain running	Fuel system
Engine starts unpredictably or starts in certain circumstances and not others	Engine compression or ignition system (check throttle position)
Engine requires many revolutions and full choke to start, but will reliably start	Fuel system (in very specific circumstances this can be caused by poor engine compression)
Engine doesn't start, but will backfire (either through carburetor or muffler)	Ignition timing
Engine easily floods	Fuel system or improper use of choke

then you should skip to the section on troubleshooting the starting system. If your engine is a manual start, or your tractor's starter is clearly spinning up your engine fine, then your problem lies in one of the remaining three areas: fuel delivery, spark, and compression.

Then conduct a fuel presence test, though the fuel system is actually less likely to be the culprit than the ignition system. The main reason we check for fuel first is convenience and simplicity. Checking for a spark is quick and easy also, but is much less certain. The spark may not occur for a whole raft of reasons that don't have to do with the problem at hand (bad ground or other factors). The fuel presence test is an either-or type of test: Either gas is present on the plug or it's not.

The process simply involves removing a spark plug or two and trying to determine if the smell of fuel is present. First make sure you do not have residual fuel smell on your hands. I have been fooled by

this before. Wash your hands or use gloves if necessary. Next, try starting the tractor with the choke on. Just several revolutions is all you need. Any more than that and you will flood the engine, creating additional problems. Immediately remove the spark plug and take a small whiff of the tip of the plug. There should be a distinct smell of gas on the plug, and in fact it may be wet with gas. You should suspect fuel delivery if you have trouble smelling the gas or it seems very faint. If you wonder if gas is present, then it probably isn't. A healthy fuel system delivering fuel to an engine while being choked will put enough gas in the combustion chamber to make its presence easy to detect on the spark plug.

The most likely cause for your engine failing to start is the ignition system. As I mentioned above, for an engine to start, a healthy spark must be present in the cylinder and it must be delivered at the right time. Checking for a healthy spark is relatively easy, but a lot less tedious and time consuming if you have a

helper. The first step is to move the tractor to a dark area, remove a spark plug, and ground the electrode or body of the spark plug to any part of the engine that is convenient. I pick a place on the engine that is without paint, and then I use sandpaper or a file to expose fresh metal. Finding or creating fresh metal is important, because choosing a weak grounding point can cause a failure to get a spark or can alter the strength and appearance of the spark. Once you have properly grounded the spark plug, have the helper try to start the tractor. If the spark plug is adequately grounded, and the tractor is in heavy shade or in a building, a healthy spark will look whitish blue and will seem strong and appear to occupy the entire gap of the spark plug. If the tractor is in a heavily lit area, then the spark may appear bluish white. A weak spark in these conditions will appear small or faint and slightly yellowish and will look like a lightning bolt between the electrodes instead of a flash that occupies the entire electrode gap.

Before you pronounce the spark missing or weak, double check your grounding procedures and be positive that you have properly grounded the spark plug. A healthy spark will also make a noticeable sound that is reminiscent of snapping your fingers, only quieter.

The twin sister to getting a good spark is making sure it occurs at the right time. Typically, timing is not something that completely changes all by itself one day. The timing usually is something that is a bit off, causing rough running, poor acceleration, and the like. If your tractor docs not start because of timing, then your engine is way out of time and you can bet something more than general wear and tear has occurred. It could be that a distributor or magneto was removed and reinstalled incorrectly, a mounting bolt came loose, one of the timing gears lost a tooth, or perhaps the drive lugs of the magneto/distributor are worn. Checking the timing is not difficult, but since the engine isn't running, you will have to time the engine in a way that may be unfamiliar to you. The method is called static timing and simply involves moving the distributor/magneto in relation to the engine so it fires at the correct point in the engine rotation. The method is outlined in the repair/restoration chapter, and I encourage you to skip ahead and static time your engine now so you can be assured that the spark is occurring in time with the engine.

The last area to check is engine compression. To fire, the engine must be capable of generating compression. While the actual lower compression limits vary with the engine, I start becoming convinced compression is the culprit when cylinder compression readings are below 60 psi cold. Some engines will start reliably with compression readings above 40 psi, but most become unreliable below 60 psi. If compression is the culprit, then I like to add a tablespoon of heavy oil to each cylinder, wait five minutes, and try to start the tractor again. The oil will often elevate compression enough to start the engine if the compression is the culprit and the compression is borderline. If the compression is horribly bad, or the fault lies elsewhere, oil won't help. Often engines with poor compression will start when they are cold and will not start when they are hot. Another clue is if the tractor starts easily in one season, but won't start easily in another season. Regardless of your success in starting a tractor with compression this low, you should realize that a tractor with compression lower than 60 psi is a candidate for a rebuild anyway, and you should consider rebuilding the tractor's engine.

Troubleshooting a stubborn engine involves examining three main systems: fuel, spark, and compression. If you can narrow it down to one system, you can then troubleshoot that particular system further in a structured, coherent way. Even if you can't narrow the problem to one system at first, the troubleshooting procedure is the same. You can go through many times correcting each problem as you identify it. Fixing each problem will then help you to pinpoint others. If

> **SAFETY NOTE**
> If you are performing this procedure on a horizontal two-cylinder John Deere, be sure the carburetor float is working perfectly! The carburetor, if leaking fuel, will fill the cylinders with this fuel. If you remove the spark plug, ground it near the spark plug hole, spin the engine and a spark occurs, the spark will ignite the fuel that will be pushed out of the spark plug hole. This is no joke! I have had several acquaintances and friends receive third-degree burns and have property destroyed because of this weakness in the horizontal engine design!

you correct a weak or nonexistent spark, and the engine still won't start, your problem is one of fuel or compression. I would say most antique tractors, especially if they were just rescued from the hedgerow, typically have several things wrong with them and you will have to troubleshoot many problems. If you are well organized and systematic in your troubleshooting, you will be able to identify all problems and bring the engine back to life.

POWER DELIVERY PROBLEMS
Engine Hesitates or Stalls at Higher RPM and Heavy Loads

This symptom often appears in unison with another symptom: rough or uneven idling. Both symptoms are usually caused by dirt or water in the fuel system partially clogging the main jet. Under idle, the stoppage lets enough fuel through to allow

Looking closely at the threaded end of this shaft, you can see the thread has become damaged. When threads become this damaged, they must be repaired before the part can be put back into use. In this circumstance, you have no choice but to replace the shaft. You can also remove the shaft and have it repaired by a machine shop. They will remove the threads using a lathe, build it back up through welding or a similar process, and cut new threads.

idling, but when fuel needs increase, the stoppage starves the engine of fuel. The engine stalls, but then fuel needs decrease, so then the engine picks back up again, only to repeat the cycle over and over again. How severe this problem is depends on the type of blockage and its extent. If the problem is water in the fuel, it will be fairly severe, but the symptoms may not be consistent. The problem may come and go as water works its way through the carburetor. Eventually, a water droplet will form at the jet large enough that the tractor will barely run. If the blockage is due to rust or dirt, then the problem may or may not be severe, but it will tend to be consistent or it may get steadily worse.

These are generalizations, but hesitation in acceleration or load is usually caused by a fuel-related problem. To see if fuel delivery is the problem, thoroughly disassemble and clean your carburetor and reassemble it. This procedure is covered under the major repairs and restoration chapter. It is very important that you are fastidious about cleaning out the jet ways and fuel passages. Now, add fuel to the carburetor bowl and install the carburetor to the tractor, but do not hook up the fuel line. If you can start the tractor and have it run smoothly before it runs out of fuel, then the problem was most assuredly with blockage or water in the fuel. Reconnect your fuel line, being sure to clean and replace the filters, strainers, and sediment bowls beforehand. Look especially for signs of water. Add de-icing solution or fuel treatment to your fuel tank at this time as well (these solutions help to homogenize the water and fuel in the tank so the water will pass easily through the carburetor).

If the problem returns after a few minutes or hours of running, the cause may still be in the fuel system. Some other causes include a poorly adjusted carburetor, a crimp in the fuel line, a leaking fuel line, a stuck carburetor float, or an obstruction in any other part of the fuel system. Typically these causes produce similar symptoms as the main jet blockage; usually they do not occur immediately upon throttling up, but take a few seconds to occur after acceleration, or placing the tractor under load. In any case, then you will need to remove your gas tank and thoroughly clean it. If the tank is rusty in the interior, I would seal the inside of the tank with a sealing compound as well.

Note that engines misfiring on one cylinder also typically exhibit this behavior, especially in acceleration. The tendency to stall or hesitate is not as strong as a fuel-related cause, however.

Engine Runs Fine except Hesitation Occurs at High RPM

The two most likely causes are both related to the points. The first is maladjusted or burnt points and the other is "point set bounce." Point set bounce occurs when the points' leaf spring is burnt, broken, or unattached. The signature symptom of both these causes is a hesitation only at the very highest engine speeds, and even then the hesitation is very rapid and fairly consistent; almost a staccato type of sound from the engine. If these are the causes, the problem usually does not change characteristics when the tractor is placed under load. The reason these two possible causes produce this symptom is because neither allows the coil to charge fully. The points, if burnt, will alter the electrical relationship with the coil, preventing the coil from producing a strong spark, especially at high engine speeds. Point set bounce results in the same thing: points that do not stay closed long enough. Checking the condition of the points is easy. The procedure is covered in the maintenance chapter.

Engine Idles Roughly or Stalls at Idle

This is most usually caused by an intake manifold leak. Very often a tractor with this problem will require that the choke stay on to keep the tractor running smoothly. Antique tractors are even more susceptible than automobiles to manifold rust-through and warping. Connecting a vacuum gauge to a port on the tractor's manifold will give you some type of indication whether manifold leakage is the problem. By the way, sometimes you will have to make this port. Usually the manifolds have a pipe plug on them somewhere. Remove this, and insert a pipe plug with a hose-barb (available at auto parts stores). You can leave this installed with a cover for future use, but I always prefer to remove it and reinsert the original pipe plug. With the tractor idling, watch the gauges for erratic, fluctuating readings. The absolute reading, for this problem, is usually not important. If you get this type of gauge behavior, finding the leak is reasonably easy. Leave the gauge installed and have a helper squirt WD-40 or similar lightweight penetrating oil in short bursts around the edges of the intake manifold and around the back side of the intake manifold while the tractor idles. Be careful! The WD-40 will burn on the exhaust portions of the manifold. Applying the fluid too liberally may cause a fire! Once WD-40 reaches the leak, the leak will suck in WD-40 instead of air. This will momentarily cause the engine to smooth out and the gauge reading to settle down. This will be very momentary so watch and listen carefully. Listening to the intake manifold will sometimes reveal the intake leak also.

Engine Shuts off When It Is Hot and Won't Restart until Cool

If the tractor stalls abruptly and unexpectedly after running for more than a few minutes, especially when operating in hot weather, the ignition coil is usually the culprit. This often is accompanied by the inability to restart the engine for a certain period of time. The coil, whether it is in a magneto or a battery coil system, usually fails completely or only fails after it heats up. How long it takes for the coil to heat up depends on a lot of factors, and even the temperature threshold at which it fails can change every time. This makes troubleshooting difficult. The best approach is to test for a spark immediately after the tractor stalls. If a spark is nonexistent, wait for a total cool-down, and then test for a spark again. If you then have a spark, your coil is at fault. Occasionally, a component called an ignition resister is not included when a conversion to a 12-volt ignition system is made. This will cause coils to fail in much the same way.

Engine Shuts Off and Won't Restart

This problem rests most likely with the ignition system. There can be many causes, and I think I have run across all of them at one point or another. In most cases, the points are involved. Either the points are burnt completely, or they are glazing and pitting due to a bad condenser. To test if the points are the cause, simply open up the point compartment of the distributor and lightly sand the points with a points file or emery cloth. Be sure to remove all traces of grit and clean the points with a business card soaked in electrical cleaner. If the engine starts, you need to diagnose the reason for the bad points. If they look like they have

Disassembling antique tractors for restoration or repair will yield high mountains of parts. Unless your memory is better than mine, it will serve you well to label and organize your parts so you can reassemble with a minimum of delay and confusion later on.

a hard glassy coating that is clear, gray, or black, or they look ridged and bumpy, then the problem is with the condenser. Replace the condenser. Clean up the points and see if the problem disappears. If the problem reappears, or the points looked fine but sanding them cleared up the problem, then the points are defective.

If working with the points was of no help, and the engine is still unable to start, check the secondary ignition wire (the wire between the distributor and magneto or coil). Testing it involves substituting a known good spark plug wire (I often use a spark plug wire from the engine itself) and trying to start the engine. Remember, if you removed a spark plug wire to test the engine, the engine will run rough if it does start. If it starts, though, you have found the problem. Do not bother checking the secondary wire with a continuity test on a multimeter. This wire can be capable of carrying the low test currents of a multimeter but be incapable of carrying the high voltage of a running engine.

Other possible causes include a bad coil, bad ignition switch, and loose connections. Loose connections can be found best by simply inspecting, loosening, cleaning, and retightening. Go through all the ignition connections, checking for wire breaks and bad connections. The coil of a battery system can be checked for absolute failure using a multimeter. With the ignition switch on, and the engine rotated so the points are closed, check for nominal voltage between the "hot" side of the coil and ground (pick any place on the frame). If you get voltage, that doesn't mean the coil is fine, but if you don't get voltage, then the coil is definitely bad. Unfortunately, testing the coils of magnetos is beyond the scope of this book. There is an excellent book called *How to Restore Tractor Magnetos* by Neil Yerigan. This, in addition to some additional information on your particular magneto, will help you properly test your magneto's coil.

Engine RPM Is Variable at the Higher Throttle Settings

The typical scenario when this problem occurs is the times the operator throttles the tractor up, and the engine can't seem to "find" the rpm it is looking for. While the tractor is not under load, it does this at only the highest throttle settings. When the tractor is under load, this problem will occur at a broader set of throttle positions, though it still only occurs at the higher settings. This problem is also known as "hunting" and is almost exclusively caused by a weak throttle spring. In effect the throttle and the governor are having a tug of war. The end result is the governor is "bouncing" against the spring, causing the variability in engine rpm. Testing whether it is the spring is easy: Simply take a length of stiff wire (paper clips are handy here) and connect the spring to the governor arm using the wire in a way that shortens the effective length of the spring. This, in turn, makes the spring tighter. If the problem is indeed with the spring, the problem should either go away or be minimized with this shorter spring. If the spring is the cause, the simplest fix is to shorten it. This fix can cause problems if you overshorten the spring, as you will lose some high-speed horsepower. Also because the spring is already worn, the problem will probably recur within a few hundred hours. I would replace the spring if possible, but you must find a spring with identical length and strength specification. This can be difficult. In the past I have also searched for a parts tractor with a good spring and used that instead.

Very rarely, the problem resides in the health of the carburetor or improper throttle linkage adjustment. If replacing the spring does not clear up the problem, you should check the throttle linkage by inspecting, cleaning, and adjusting it according to your tractor's service manual. If the problem continues occurring, then fully overhaul your carburetor. Blockage at the main jet and improper high-speed idle adjustment can cause this problem too. If the carburetor is in need of general cleaning and adjustment, this problem can occur.

When Engaging the Clutch, or Exposing the Tractor to Varying Loads, Engine Speed Increases without a Corresponding Increase in Power

Let me explain the symptoms of this first in a little more detail: In this scenario, a rapid increase in power requirements such as engaging the clutch or hitting a thick section of growth with a brush hog, will cause the engine to increase its rpm without the tractor or implement receiving any of the extra power. In addition, the engine speed may fluctuate for a while and then settle down, or it may fluctuate continuously if the load is particularly heavy. This is usually caused by clutches that are slipping because they are worn and need replacing, but it can be caused by clutches that have been exposed to grease or oil, or clutch linkage that is poorly adjusted. There typically is no noise associated with this problem and repairing the problem almost always means replacing the friction plate(s) within the clutch assembly. Check the linkage first to see if that is the problem. Typically a clutch, whether it is a hand clutch or foot clutch, will have some amount of free play in the linkage. No free play in the travel of these clutch linkages can cause clutches to slip by leaving the clutch and throw-out bearing in a state of preload. You should adjust the linkages first before doing any other trouble-shooting. On most tractors, repair boils down to replacing the friction plate. On some tractors, such as John Deeres with hand clutches, the clutch itself can be adjusted to account for wear. Your tractor's service manual will outline adjusting the clutch.

CLUTCH-RELATED COMPLAINTS

Troubleshooting a clutch, with the exception of a few circumstances, can be difficult for owners of antique tractors. Many of the noises they make when they go bad are often subtle and difficult to discern. Often the sound seems to be coming from the engine. Even when the sound clearly emanates from the clutch, is it the clutch, throw-out bearing, or possibly a pilot bushing or bearing? The clutch is never exposed in full view; many antique tractors have many different kinds of clutches, and some models have more than one clutch. The auxiliary clutches are for the engagement of the PTO or belt pulley. Add to all this the fact that some are hand clutches and others are foot clutches, and you understand the challenge tractor clutches present. Troubleshooting clutch problems means truly understanding the problem, the nature of the sounds it makes, and the situation under which it occurs. I will attempt to steer you in the correct direction

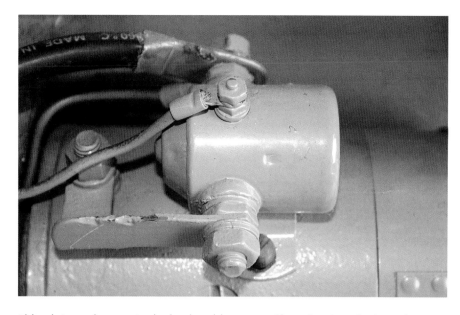

This picture shows a typical solenoid on an antique tractor starter. The larger heavy cable provides the power for the starter, while the smaller cable comes from the start button or ignition switch. This smaller cable operates a switch inside the solenoid that makes or breaks the connection between the larger cable and the starter. The flat bar coming from the near side of the solenoid carries the current from the solenoid to the starter.

and at least cover some of the troubleshooting techniques.

Listening to sounds and determining their origin is tough business for the inexperienced ear. The sounds can be quite varied, and identical problems on two different tractors can sound quite a bit different. The symptoms of a malfunctioning clutch typically fall into one of three categories: noises that occur while the clutch is engaged and the tractor is operating, noises that only occur when the pedal is depressed, and stuck clutches.

The first scenario, noises that continuously occur while operating, are usually associated with a worn or glazed friction plate. In that event, the noise is produced by the clutch slipping. The noise and the amount of slipping will vary according to the load on the

tractor. Another possibility is the pilot bushing or bearing. In this case, the noise will vary with engine rpm, and in the case of bearings may slightly worsen when the tractor is under load, but not necessarily change with the loads. Noise that occurs when the pedal is pressed is usually associated with a bad throw-out bearing. The throw-out bearing is a thrust bearing that activates the clutch release mechanism. Other possibilities include a warped pressure plate that disengages the clutch unevenly, causing occasional or constant light contact with the friction plate. Noises associated with warped pressure plates are a chattering noise or a whirring. Warped pressure plates are also marked by uneven engagement of the clutch and noises during engagement.

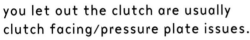

Antique Tractor Tip

Diagnosing Clutch Noises
With clutch noises, constant noise is usually a pilot bearing/bushing, noise when you press the pedal down is usually a throw-out bearing, and noises that occurs when you let out the clutch are usually clutch facing/pressure plate issues.

One other problem you may run into with an antique tractor is that of a stuck clutch. Stuck clutches are just as they sound: The clutch will not disengage. The clutch acts as if the pedal is not pressed, even when it is. The clutch is rusted into the engaged position in all likelihood. "Un-sticking" a clutch can be accomplished through some very ingenious means, all of which involve running into immovable objects at slow speeds on purpose, but I don't recommend them. The proper and safe repair is to remove the clutch and then clean and install. Usually the friction plate has to be replaced.

The bottom line for all of these problems is that you must expose and inspect the clutch. Some inspection can be done through inspection holes or by removing an electric starter and slowly rotating the engine by hand; but these inspection methods are typically only good for initial assessments and should not replace a thorough visual inspection of the entire clutch. Getting to the clutch is a lot of work, and you should refer to the section on repairing and rebuilding the clutch for more information.

STARTING SYSTEM

Ever since the advent of electric starting in tractors around the 1930s, tractors have been easier and more convenient to start. Of course, that is also just one more system to maintain, and to repair when problems occur; but fortunately, the starting circuits of antique tractors are robust and effective. The 6-volt electrical system most antique tractors have causes the starter to turn the engine roughly half as fast as the 12-volt systems we are more accustomed to. This leads many people to incorrectly assume that the slow cranking speed is the culprit of their engine not starting. This is simply false, as a 6-volt starter will start any tractor a 12-volt will start. The key is having an engine in good mechanical condition. The switch in the 1950s to 12 volts was made for several reasons: primarily to decrease the amperage flow through the starting circuit and to provide brighter lighting. It had nothing to do with a 6-volt system's inability to adequately start tractor engines. If your engine isn't starting but your starter is turning the engine, your starting problems are due to something else.

Does the starter spin quickly, but it doesn't engage the engine? If so, the problem is with the bendix gear of the starter. Removing the starter and cleaning and lubricating the bendix starter drive will fix the problem. If the starter makes a strong humming noise but doesn't turn, then the starter most likely has gone bad, but the problem may be the engine is frozen or the starter is trying to move the tractor forward but can't. These last two possibilities are easy to eliminate. The first, trying but failing to move the tractor forward, is usually caused by a clutch that is rusted into the engaged position and the tractor is in gear. This isn't common, because most healthy starters will also lurch the tractor forward, but a weak starter, battery, heavy tractor, parking brakes, and so on, can all create an insurmountable force for the starter, causing a hum. The second, a frozen engine, is also simple to diagnose. With the tractor in neutral, try to spin the engine by hand. You can do this with a hand crank, or spinning the engine by pulling a belt, or on older tractors or John Deeres, by spinning the flywheel. If the engine can be made to turn without a lot of effort, then the engine is loose and you can rest assured your starting problems lie with your electric starting system.

The first thing to do is to make sure the battery isn't dead. A simple test with the lights of the tractor, or with a voltmeter or hygrometer will tell you if the battery is weak or strong. Next, see if the switch on the starter (not the ignition switch, the switch on the starter called a solenoid) is okay. The easiest way to see if the solenoid is okay is to listen to it. An electric solenoid should make a clicking sound and a manual starter switch will make a slight zapping sound when it engages. Unfortunately, it is sometimes difficult to hear from the operator's platform, and manual switches in particular are very quiet. Therefore, the surest way to determine if the solenoid is okay is to measure the voltage at the battery during starting. If the voltage drops at the battery when the starter is engaged, then the problem is with the starter.

If you don't hear the click or see the voltage drop, the problem is with the solenoid or the cable from the battery to the solenoid, or the cable from the battery to ground. The first culprit is almost always the wiring. Let me dispel the myth that a simple continuity or resistance test with the volt meter is useful in checking these cables. A cable that carries a minimal voltage for a continuity check will not necessarily handle the several hundred amps that a starter draws. There is no substitute for removing the connections, cleaning them, and seeing if the problem corrects itself; tests with a multimeter won't tell you. Continuity tests are helpful for the low current, low voltage connections, and to determine the condition of the ground. But for large current or

Antique Tractor Tip

Starter Savior
Starters that sound slow and weak are usually fine, but act this way because they are fed electricity through old, weak, or broken cables with poor connections to the battery, starter, and/or ground. Double check all cables and connections before suspecting your starter.

high-voltage wiring, continuity checks are useless unless you suspect an actual break in a wire and you cannot see the wire to tell.

To check connections and cables, disconnect the ground terminal of the battery. The vast majority of antique tractors were positive ground, but many have been converted to the modern negative ground standard. Simply look to see which terminal is connected to the chassis or frame of the tractor. The terminal this cable is connected to is the ground pole. For example, if the positive terminal of the battery is connected to the torque tube of the tractor, then the electrical system is a positive ground system. Disconnect the ground terminal at the battery. Make sure the cable is thick (at least 0 gauge) and the connections are clean and tight. Check the ground point for good contact. Rust or a fresh coat of paint can keep the battery from making a good ground. The area where the ground cable connects to the tractor must be clean and

bare. Leave the ground cable disconnected while you check the rest of the system. Next, check the cable and connections of the cable carrying the current to the battery. Again, all connections must be clean and tight. The cable must be thick for 6-volt systems and as short as possible is best.

If your tractor has an electric solenoid, check to make sure the wire carrying the current to the solenoid is in good condition and the connections are clean and tight. The solenoid is activated by some type of push button or by the ignition switch. Make sure this switch is healthy. This is one case where a continuity and or resistance test with a multimeter will help. The last connection to check is the connection between the solenoid and the battery. Manual switches usually contact the starter armature directly, but the electric solenoids usually have some type of copper bus that connects them to the armature. Make sure these connections are clean and tight. If after making these checks the starter does not spin up and the solenoid

Antique Tractor Tip

Charging System Tool
A good multimeter is a must for troubleshooting charging system complaints.

doesn't make an audible click or cause the starter to draw some type of current, then your solenoid is suspect. Remove the starter solenoid and replace it.

CHARGING SYSTEM COMPLAINTS

The charging system is made up of four primary components: the generator (or alternator, a common retrofit), the voltage regulator (or cutout), the battery, and the wiring. Troubleshooting requires a reasonably thorough understanding of the theory of how your tractor should produce, use, and store electricity. The generator provides the electricity for the electrical system and in the process it keeps the battery charged. An alternator does the same thing, except in a slightly different way. Electricity is produced by either system. The electricity passes through a separate device called a voltage regulator or cutout (different from voltage regulator but performs a similar function) and then on to the ignition system, lights, and battery. In the process the ammeter tells you if the generator is providing more than enough electricity (by reading in the charging areas) or if the battery is supplying some (in which case the meter will read in the discharging location).

Battery Is Not Staying Charged

The first test to determine if the battery is at fault or the generator is at fault is a charging system voltage test. Simply run the engine at high speed and measure the voltage across the battery terminals. The voltage should read well above the nominal voltage of the electrical system: 6-volt electrical systems should read 6.5 to 7.0 volts, 12.0-volt systems should read 12.5 to 13.5 volts. If you see these types of voltages, then your charging system is charging and your battery is bad. If you do not read voltages in this range, but instead read less nominal voltage, then any other charging system component can be at fault. The first thing to check is the cutout or voltage regulator (hereafter, both are called the regulator). Test the voltage at the regulator by clipping a lead from a multimeter to the base frame of the voltage regulator and the second lead to the armature terminal of the voltage regulator. If your system is a positive-ground electrical system, clip the positive lead of the multimeter to the base and the negative lead to the generator terminal. Vice versa for negative-ground systems. With the engine running at or near full speed and the lights on, read the multimeter. You should see a voltage reading at or near the 7 or 13 volts mentioned above. If the reading is highly erratic, the regulator is bad. If the reading is steady and low, or 0, then either the generator or the voltage regulator is bad. If the system is reading an output voltage at or slightly over the nominal

voltage of the electrical system, the charging problem is in the wiring or the ammeter. Check the wiring connecting the battery terminal of the regulator to the ammeter. Then check the wiring from the ammeter to the common post on the starter. Any of the connections or the ammeter will be the culprit.

Battery Is Being Overcharged

This problem is marked by excessive loss of battery electrolyte. When the charging system is overcharging the battery, there is usually just one of three reasons. One, the generator has a three-brush system that is incorrectly adjusted. Two, the current-limiting circuitry of the voltage regulator (if present) is inoperative. Three, the field terminal of the generator is grounded to the frame. If the tractor has a regulator, then suspect that unit first. If the tractor does not have a regulator, and instead has only a simple generator with a cutout, then the problem is most likely that the generator has a third brush that needs ad-justing. Otherwise, somehow the field terminal of the generator is grounded to the frame. This was commonly done to work around the notoriously unreliable cutout control devices that preceded the more modern two-contact voltage regulators. The side effect of this "fix" is that grounding the field terminal causes the generator to put out as much current as possible, which will "cook" a battery, especially if the generator is of the three-brush type.

Typically, though, the regulator is at fault. Replacement is usually called for, but you can sometimes repair these yourself. After disconnecting the battery ground, remove and disassemble the regulator. Some of these units are riveted together and cannot be disassembled without destruction. They simply must be replaced. Most, fortunately, are screwed or bolted together. Upon disassembly, you will see one, two, or three sets of coils of wires. Above or below these coils are contacts. Clean these contacts with a clean business card soaked with electrical cleaner and then burnish the points with a point burnishing tool or a very fine grade of emery cloth. Move the contact levers several times, checking for smooth, bind-free operation. Reclean the contacts, reassemble, and check the operation of the generator by reading its output, using the ammeter of the tractor or a hand-held ammeter. Is the ammeter still showing full charging after several minutes of operation? Then possibly the third brush of a three-brush system is improperly adjusted. Again it is outside the scope of this book to discuss how to adjust the third brush of a three-brush system, though it is not outside the abilities of many people. I suggest further reading and research if you want to try to adjust this yourself.

The Battery Discharges When Using the Lights

While most generators will not supply enough current to light the lamps and keep the battery charged when the tractor is idling, the generator should be able to do this when the engine is running above idle speed. To diagnose this problem, you should realize that it is closely related to the problem above in which the battery overcharges. The same sets of causes are probably at work and much of the same troubleshooting should be done. The likely candidates are in reverse order, though—a loose generator belt or improper adjustment of the third brush of the generator. The regulator is less likely to be the cause. If the regulator were bad, it would have to be partially operational since it can keep the battery charged without the lights. After making sure the generator belt is tight and in good condition, check the voltages at the battery and the regulator as described in the last section. If the voltage is okay, I would try to adjust the third brush. Move it away from the output brush, reassemble, reconnect, and start the tractor, watching the ammeter on the tractor. If the problem continues, service the regulator as described above. If this doesn't repair the problem, the generator may be at fault and a shop will have to bench test the generator and the regulator together.

There are many excellent generator and starter shops that will rebuild your tractor's generator and voltage regulator for a reasonable fee. This fee includes diagnosing which unit is bad.

TRANSMISSION COMPLAINTS

The majority of transmission complaints boil down to the inability to change gears or reach certain gears and noises from the transmission. The inability to change gears may be clutch related. Do the gears grind when you try to change gears, even though you have the clutch pedal pressed in? If so, the problem is with the clutch. If you cannot reach some or all of the gears in the transmission, especially if you feel like the gear shift is operating "outside" of its normal pattern, then the gearshift has jumped its fork or rails. This usually is an easy fix that involves removing the cover through which the shifter extends. Then adjust the forks or rails to put the shifter back into its tracks. The fundamental problem that allows this is excessive wear. A permanent fix will involve disassembly and having a machine shop build the worn component

Antique Tractor Tip

Clicks and Knocks
A periodic low "clicking" or "knocking" noise coming from the rim/wheel/tire area is usually a loose or missing rim-to-wheel mounting bolt.

Before power steering was widely available on antique tractors, aftermarket systems were available. Here is a belt-driven example on a tractor owned by Gene Dotson. These systems are tough to come by and are even tougher to find replacement parts, but modern hydraulic shops may be able to help with some repairs and parts.

back up. Typically, this problem happens because the end of the shifter is worn. While you have it disassembled, I suggest building up the end of the shifter by enlarging the end through welding. This will prevent it from jumping its notches in the rails or forks. On some tractors the rails and forks can be difficult to access, but the process is the same. Noises from the transmission should be grouped into one of two categories: noises that happen from a specific gear and noises that happen all the time. If the noise occurs in a specific gear, and the noise is an even whirring or whining noise, then the gear is worn. If the tractor clearly operates fine with the noise, most antique tractor owners just try to live with it. If it is very loud or is concerning you, rebuilding the transmission is the only option. By the way, the high-speed road gears of these tractors often make whirring and whining noises and are a reflection of the higher transmission speeds and design clearances. These should not be considered a problem, but a feature of antique tractors. The noises from these road gears should not be overly loud, though, and you should investigate any road speed noise that is unduly loud. If the noise from a specific gear is a "hammering" type of noise, the gear is missing teeth, and the transmission should be immediately rebuilt. If rebuilding isn't feasible, and you can avoid that gear, the tractor should be fine to use.

Noises that occur independent of specific gears are almost always related to bearings. Exceptions are tractors such as the Ford 8N that has the hydraulic pump in the transmission cavity, as the pump could be causing the noise. In addition, these noises may actually be coming from the PTO or belt pulley gearing if those units are engaged. Disengage these units before troubleshooting the transmission noises to be certain they are not the cause. Other possibilities include pilot bearing and synchronization gearing. With the exception of the transmission input shaft bearing, which can be reached if the tractor is split in half at the transmission, all the components I mentioned can only be corrected through transmission removal and disassembly.

HYDRAULIC COMPLAINTS

The hydraulics of the tractor are difficult to diagnose without proper tools and training, but some common complaints lend themselves to owner-operator troubleshooting. Before you do anything at all, I insist that you change your hydraulic oil, being sure to use compatible hydraulic fluid. The vast majority of hydraulic system complaints I have run across are due to improper fluid, old waterlogged fluid (particularly common in transmission case hydraulic systems like the Ford and Ferguson tractors of the N and TX series), or low fluid levels. Now that you have done this, we can begin.

The most common complaint is that the hydraulics system is not lifting the loads or not lifting heavy loads as it should. The other common complaint is that it will not hold the load up for any reasonable amount of time. Other less common complaints are pump noise, jerky movement, and improper draft control behavior. The hydraulics system is best tested when the tractor has been fully warmed up and implements that are typical of your uses have been attached. The trick to troubleshooting hydraulics is understanding how it works. The pump creates the pressure needed to raise the implements, but the control valves (which are the valves you operate) are the components that build, maintain, or remove that pressure. The devices that actually lift and lower the implements are nothing more than pistons within tubes reacting to the force of the hydraulic pressure that the control valves apply. If your tractor has draft control, then there is another device that reacts to the implement. This draft-sensing mechanism simply operates as an automatic control valve and is basically no different than the control valve you operate. Once you know the basics, troubleshooting involves keeping in mind how the system works and logically following through the system identifying possible causes.

We will use the complaint of "My system cannot lift an implement" as an example of what I am talking about. This problem is completely different from a lift that will not operate at all. This complaint assumes the lift operates normally with no implement. To lift an implement, the lifting apparatus must receive adequate hydraulic pressure. The lifting apparatus, regardless of type, is just one or more pistons in a tube(s). These pistons therefore, must not be moving at all if the lift doesn't work. Since they are not moving, they are stuck (highly unlikely) or they are not receiving hydraulic fluid, or pressure, or both. Low fluid levels are easy to check, so make sure now that the fluid level in your hydraulic reservoir is topped off. Use your owner's manual to find out how to do this on your particular tractor.

Now we are sure we have fluid, but if it still doesn't work, we know the problem is inadequate pressure. To check the pressure, use a special hydraulic gauge. Every manufacturer has located somewhere in the hydraulic system a port, or hole, in which you attach a hydraulic tester. This can be rented from most tool rental yards. This device will tell you how much pressure the system is generating. Usually the port is after the control valve in the hydraulic fluid loop, so you must open the control valve to test. Read the reading and compare against your service manual. If the reading is low or nonexistent, then the problem is with your pump, cylinder, or control valve. If the reading is normal, the piston(s) that actuate(s) the lift is stuck or the implement or lifting apparatus is pinned against something immovable. If the fluid pressure is low, then one of several things can be at fault, and generalizing for all of the different types of hydraulic systems is difficult. Rest assured, however, that the fix will involve considerable work to the lifting pistons, control valve, pump, or all three. Consult your tractor's manual for additional troubleshooting.

Hydraulics that lift jerkily are filled either with hydraulic oil that is waterlogged or with an oil that was detergent based. Detergent-based oils foam under the pressures of hydraulic systems, leading to symptoms such as this. Hydraulics that will not keep an implement lifted in the air have leaking O seals either at the lifting

Antique Tractor Tip

Hydraulic Hints
Hydraulic systems with jerky, spongy movements usually have detergent-based oils in them or contain excessive amounts of water.

cylinders or the control valves. Weak pressure relief valves and improper hydraulic fluid can cause this. Consult your tractor's service manual or a local hydraulic repair shop for other complaints.

Conclusion

Figuring out what ails your tractor is an exercise in logic. It can be frustrating, and there are many situations that can cause even the most experienced mechanic great trouble. I do want every reader of this book to realize that these circumstances are few and far between, and the vast majority of problems can be discerned. Reading additional books on the theories and design of engines, transmissions, electrical components, hydraulics, and the like is an important first step in being able to troubleshoot quickly and well. Taking every opportunity to work on broken tractors is another excellent teaching tool. Joining antique tractor clubs and attending antique tractor shows will go a long way toward training you in the science, and art, of troubleshooting the antique tractor.

CHAPTER 13
Engine Repair and Restoration

Seen here are the main differences between gasoline and diesel engines. First, you may notice the starter is larger. Diesel engines have a much higher compression ratio, are harder to turn over, and therefore require sturdier starts. Above that are the fuel filter and water separator. Fuel quality is important in diesel engines so there is additional fuel conditioning. To the right of the filter and separator is the injection pump for injecting fuel into the cylinder. Diesel engines are ignited by the high temperatures created by the high compression and are not spark ignited. Fuel must be injected into the cylinder in a fine mist to improve burn rates and speed the burn rates of the fuel.

Sooner or later virtually every antique tractor engine will need some level of overhaul, or a total rebuild. Fortunately, because these engines are relatively simple, most repairs, even a top-to-bottom rebuild, are within the competence of the hobby mechanic. The important thing with engine work is to proceed carefully and methodically. Not only must you keep track of all the parts, you must also be sure each part that is reused is returned to the precise location from which it came—even if it is one of several apparently identical parts. Stuck parts need to be removed properly, to avoid damaging other parts or injuring yourself. And you must be a careful observer, so you can detect which parts need to be repaired or replaced, and which can be reused.

In this chapter I will discuss some preliminary considerations, and then cover the phases of the rebuilding process. I provide repair and safety tips that I have learned and also note some unusual approaches I have not tried. In addition, I mention some dangerous techniques that you should avoid. Bringing an abandoned engine back to life, or restoring health to an ailing engine, may be the most satisfying part of the restoration process. Let's get that old powerplant back into shape.

General

ENGINE OVERHAUL: STRATEGIES AND DECISION MAKING, INTRODUCTION AND PLANNING

Usually there is some obvious reason that an engine overhaul should be done. For example, an antique tractor may sit unused in a field for years before someone comes along and rescues it for a collection or to use. Having sat in the elements, the engine has often been infiltrated with water, causing it to seize up. Often the engine was in such poor condition when it was parked that even if the engine isn't stuck, starting it without significant work is impossible. Many times a serious problem developed, such as a broken piston, cracked tappet, or warped or cracked engine block or head. Under these circumstances there is no alternative to an engine rebuild (unless, of course, you have a replacement engine on hand).

Even more common than these doomsday scenarios is the antique tractor that will run and start, even after sitting for years, but has poor performance characteristics—low power, poor starting (especially when hot), excessive oil consumption, or very low oil pressure. There may be other telltale signs of an impending rebuild, such as engine knock, excessive blow-by from the crankcase breather, and so on. With these engines, it is often not clear when or if the engine needs a rebuild. If it

Antique Tractor Tip

Easy Diagnostics
If at all possible, try to get an engine running before you rebuild it, even if it runs poorly. This will give you the opportunity to learn all about the weaknesses and problems of the tractor before investing a whole lot of time and money.

runs, it doesn't need a rebuild, right? Yes and no: The answer depends on how you use it, your tolerance for a poorly performing engine, and your willingness to expose yourself to the risk of running a poor engine longer, maybe causing additional engine problems.

If you use the tractor infrequently and it starts reasonably easy, then maybe putting off a rebuild is a legitimate strategy. If you use the tractor frequently, or if it is critical that the tractor perform its chores completely when asked to, then rebuilding the engine becomes sound strategy. My motto is: When in doubt, rebuild the engine. Tractor engines are nowhere near as difficult and time consuming to rebuild as automobile engines are, and the parts costs are not astronomical. Since the engines are easy for the owner to rebuild, rebuilding it yourself is feasible, saving you the largest part of the cost, labor.

Rebuilding a tractor engine is straightforward, and planning it isn't terribly difficult. The largest issues to address involve how you will handle large and heavy

sections of a disassembled tractor, and where you will be performing the rebuild. Many tractors, such as John Deeres and Farmalls, require splitting the tractor at the transmission to perform significant work such as crankshaft repair. Other makes, such as channel frame Massey-Harrises and many Allis-Chalmerses, can be left intact when the entire engine is pulled. The other major issue, where to perform the rebuild, is important. The area should be covered and the rebuild performed on a dust-free floor such as a concrete slab. If you must use a temporary shop, make sure you have created an environment that has fire safety devices, easy tool access, and plenty of lighting. If you require heat in your temporary shop, make sure you use a heat source consistent with your use and ventilation of flammable shop cleaners and supplies.

Can You Get It Running?

Before you get out the wrenches and start tearing down the engine, I promise you can save yourself a bunch of time and trouble if you first try to see if you can get the

Antique Tractor Tip

Rebuilding Philosophy
Whether you completely remanufacture an engine or simply refurbish it is a matter that has more to do with you and your goals than any standard of technical right or wrong. Both approaches work fine in different circumstances.

engine running if it isn't running already. The reason for doing this is that a running engine will tell you about all its weaknesses. If you have not heard the engine running before you start a rebuild, then you may complete the rebuild only to find the generator is shot, the oil pump doesn't develop oil pressure, the carburetor bushing leaks air, or the cylinder head is warped and oil mixes with the water, or the like. Getting the engine running may show you that the engine's condition isn't as poor as you had guessed, allowing you to postpone or cancel the rebuild. This is especially important for engines that may be only lightly seized up when you get them. Often, the owners get the engines freed up only to find that they run pretty well and don't really need a rebuild, or the rebuild can be postponed.

A running engine can also help you find all the other faults in the tractor: Bad hydraulics and broken transmission gear teeth will become obvious now that the engine and tractor are running. Maybe there are so many other problems with the tractor that rebuilding the engine isn't worth it. If you can make it

run, you'll have a much better idea of what sort of machine you're dealing with and what lies ahead.

Determine Need for a Rebuild

To understand your engine better, and to help you in your decision-making processes, you should take several vital signs of your engine, if possible. First, take a compression test. The compression test will tell you the extent to which your engine can generate all the power it was designed to deliver. This test is covered in the maintenance section. The service manuals for your tractor will tell you what the readings should be. Without such guidance, I use a system that looks at variance between cylinders. As an engine wears, the difference in compression between cylinders becomes more pronounced. If the difference between the highest and lowest cylinder is more than 10 percent of the highest cylinder's reading, then a rebuild may be in the cards. More than 15 percent and I would recommend a rebuild. Any difference greater than 20 percent indicates an engine begging for a rebuild, and you

would be smart to oblige it. The system of variance is useful even if you know the factory specifications for compression. If all cylinders are above minimum specifications but the variance between cylinders is over 20 percent, then you should still rebuild the engine.

Oil pressure that is constantly very low and cannot be adjusted higher is another sign that you should rebuild the engine. Heavy blow-by or thick black oily exhaust indicates at least a refurbishment is in order. It is possible to have good compression and heavy, black, oily exhaust. The oil control ring can be broken or worn. Failure to correct this can cause premature glazing of the cylinders, causing compression loss. There are many reasons to rebuild an engine, and any of these reasons or any combination can be justification enough.

Rebuild, Refurbish, and Remanufacture

There are a lot of terms out there for sprucing the engine up. Throughout the book I use the generic terms of rebuild or overhaul, even though there are no clearly agreed-upon definitions as to what each means. I use either of those two words to imply any of the three following terms: refurbish, restore, and remanufacture. Here are the definitions to these three terms:

Refurbish: This means that the engine was repaired to the point that it can now generate compression readings that are within factory tolerance and will generate oil pressure readings within factory specifications. Repairs that clearly are called for are addressed as well. A complete ignition tune-up is also done. This almost never means inspecting auxiliary systems, though, such as the governor and the valvetrain, unless these systems were identified as needing work beforehand.

Restore: This means that the engine has all major components and they have been removed, disassembled, cleaned, adjusted, repaired, and assembled as needed. Some components, if within wear tolerances, were reused. All components and assemblies were cleaned and inspected, though.

Remanufacture: This means that every part of the engine was made "like new." Every part of the engine was inspected, thoroughly cleaned, and if the part was considered a wearable part, it was replaced even if it was still within wear specifications. All other parts were inspected for excessive wear. For example, all bushings, which are wearable parts, were replaced, but timing gears were simply inspected for excessive wear. The crank was turned; pistons, rings, and sleeves were new or cylinders bored, and so on. This includes all auxiliary engine systems as well, including the generator, carburetor, water pump, and so on. In addition, all machined surfaces were milled and all spinning parts, such as the flywheel, were balanced.

As you can tell, the difference between the methods has more to do with approach and thoroughness than with any mechanical or

Here is an example of a sleeved engine. The cylinder block is cast in such a way that long steel cylinders could be pressed down into the block. While many industrial engines are still made as sleeved engines, most engine cylinders are machined into the block material and are not removable.

technical criterion. Does this imply that remanufacturing is "better" than refurbishing? Of course not. It all has to do with the condition your engine is in to start with and what kind of performance and reliability you are looking for when you are finished. Every approach and method has its place, and I encourage you to decide for yourself exactly to what extent you will rework the engine.

Cost Estimates

How much will it cost? That is probably an impossible question to answer. How much you need to pay depends on how much you need to do, as well as on parts prices and availability (do you have to special order parts at extra cost? have them manufactured? and so on). A simple refurbishment that you do on an engine that was only suffering from poor compression could cost as

little as $500 in parts. An engine that needed to be completely remanufactured would cost over $2,000 depending on parts costs and any special repairs or services your engine may need. The prices vary widely, so you will have to do some planning and troubleshooting and research on your parts to find out how much the overhaul will cost.

Finding Reputable Shops

Even if you perform the rebuild yourself, you will find that some of the work will have to be subcontracted out to various shops. For example, the head and block may have to have the mating surfaces machined true or you may need valve seats pressed into place. Unless you have a fully equipped machine shop in your garage, these services will have to be performed elsewhere. Probably the best way to find a good shop is to ask the tractor dealerships whom they use. Most dealerships will be glad to provide a reference. Auto mechanics are also a good source of this information. When all else fails, look for ratings and certifications from professional organizations. There are several organizations that accredit and approve machine shops and mechanics. While certificates and training diplomas don't guarantee a perfect job, the odds of obtaining faulty service and parts are lower in rated and certified shops. Your ability to obtain satisfaction if something does go wrong is also more likely with these shops.

Disassembly

When a rebuild is called for, move the tractor to its final place, and then disassemble the engine as completely as possible while it is still on the tractor. As I mentioned earlier, on some tractors the engine is a structural part of the tractor, and removing it requires supporting the front end and the rear of the tractor. Other engines can be removed without splitting the tractor like this. If the tractor engine is seized, tearing it down in place will help you ascertain why the engine is frozen. Since many of the

nondestructive methods of freeing up these engines can take days or weeks (and even months if you are that patient) to work, having the engine in the tractor where the tractor or the starter can help may be a blessing. In other words, removing the engine prematurely almost never has any benefits and often has some drawbacks. Be sure you need to take the engine out before you actually do.

FREEING UP A STUCK ENGINE

So your engine is stuck and you are wondering how to unstick it. There are as many methods for freeing up as there are people who have freed up an engine. Most of the time being successful at it means being patient and using lots of light oil or penetrating fluid. There are all sorts of methods to try to speed things up: Some are ingenious, others are just plain strange, and some are even dangerous. Here is a listing of all the methods I have heard people try, and my comments on each one. It is important not to exert too much pressure on any one piston if you cannot or will not disconnect the connecting rods from the crankshaft. If the connecting rods are still connected, and you press down on one piston you will free that piston, but because other pistons are still connected and stuck, they will keep the crankshaft from turning. This means that all you will do by exerting tremendous force on one piston is bend a connecting rod or break a crankshaft.

Pulling the tractor in gear: This method can cause tremendous damage, primarily bending the

connecting rods (and parts of the valvetrain if the engine does come free but the valves stay stuck). Having said this, I will say this method is a good one to try first if you are very careful and tow the tractor using very slow speeds and do so over unpaved ground. The soft ground will let the wheels skid instead of causing damage to the engine if the engine is too stuck to come free.

Leverage on the flywheel or crank pulley: To use this method, simply find a way to attach a very long and very strong lever to your engine. Some people make a special adapter to fit their flywheel or crankshaft pulley. You can use a socket wrench on the crankshaft pulley nut, but too much force may strip the threads of the nut or crankshaft and it will more than likely make removing the nut nearly impossible.

Inverted hydraulic ram: This method sets up a rig whereby a hydraulic ram pushes a piston down. This method may cause severe damage to the piston if you are not careful to distribute the force evenly over the entire piston. This method is usually more trouble to set up than it is worth.

Penetrating oil, patience, and pounding: the most likely to work in most circumstances. Liberally and regularly apply oil to the cylinder walls above the piston, and pound on the top of the piston using a piece of stout oak and a small sledge hammer. Then be patient for a few days, then repeat.

"Super sauce," patience, and pounding: This is just a variation on the above, substituting penetrating fluid for some concoction "guaran-teed to work" every time. It's amazing the number of weird things people use to try to free up a stuck engine! I have heard everything from water to Coca-Cola to diesel fluid to brake fluid and everything in between. Some seem reasonably sound (diesel fuel), others border on the bizarre. If it trips your trigger, and you think it might work, go ahead. I personally wouldn't put Coca-Cola in my engine.

Grease fitting in the spark plug hole: This method uses hydraulic force to push the piston down. You can only do this to cylinders that have both valves closed. Simply construct a fitting that will screw into the spark plug hole of the engine and accept a grease fitting on the tip. This method is always messy and usually blows a head gasket before it will free a stuck engine. It will work in my experience if the engine is only lightly stuck. Takes a lot of grease.

Air pressure through the spark plug hole: NEVER DO THIS! While this method may work, there is a risk of violently blowing out the air fitting from the head. I consider the risk of damage and injury too high to use this method. Use another method instead.

Dry ice placed on top of the piston: I have never used this method, but the idea is that the dry ice will contract the piston enough to loosen it from the walls. Since the rings are usually the part actually stuck to the wall and not the piston, I cannot see this method being successful. If galling and rust have formed between the

This engine is an example of a dry sleeve engine. Around the left-hand cylinder a shiny ring is seen around the circumference of the bore. That ring you see is the top of the sleeve. The right-hand cylinder sleeve doesn't show that ring because the sleeve has been removed. These sleeves are called dry because the sleeves are not in contact with the cooling water circulating in the block.

piston and the cylinder walls, this may be a good method to start with and then use another method to loosen the rings.

Cracking the tops of the pistons out: This method works every time and ruins the pistons every time. If replacement pistons are plentiful and inexpensive, this may be the best method to use. Make sure new pistons are obtainable before doing this. Another plus for this method is the fact that most of the time new pistons are required in an engine rebuild, and the pistons you so carefully removed from your stuck engine using other methods may not even be usable! Usually it is best to try to remove the pistons intact, though. Sometimes it is not possible or is way too time consuming. Simply use a hand sledge and a steel rod. From the

bottom of the engine, place the rod on the underside of the top of the piston, being very careful to stay away from the connecting rod and wrist pin area. Hit the rod with the sledge very sharply. The tops of the pistons will come right out. This also loosens the rings from the side of the wall as well and the piston will slide right out.

Here are some methods to stay away from: exploding any substance in the cylinder. Never under any circumstances should you explode any fluid inside the engine. The risk of damage or injury is too great. Another technique to avoid is placing hot coals on top of the piston. I have seen this done for some inexplicable reason (the person claimed the expansion would loosen things up). Again, the hazards of this approach far outweigh any potential benefit.

For some reason freeing a stuck engine brings out the witchcraft in people and so much of it is pure foolishness. The thing to remember is that the rings are stuck to the cylinder wall most of the time, not the pistons! Do things to free the rings, not the pistons. Also remember that chemicals, heat, and pressure are dangerous in various combinations. One story I heard had a fellow soaking the pistons in diesel fuel. The engine never would free up, so he put the head and the gasket back on, and applied pressure to one cylinder using air pressure. Diesel is combustible, and as you might imagine, the rapid increase in pressure caused the diesel to begin to burn, bending his connecting rod. While it is unlikely that this story is true (diesel requires several hundred psi to burn), it does illustrate the need to be mindful of everything you are doing and the possible consequences.

Last Notes on Disassembly

Unless you have a better memory than I (which is likely, since mine is poor), keeping parts in labeled plastic "locking" bags is a really handy way to keep parts organized and keep them from getting lost or dirty. I learned a long time ago that those parts that I was sure I was putting in a good place were indeed the ones I couldn't find. Bagging and labeling parts is especially necessary for very similar parts. Shims occasionally look like washers, but since the thickness is critical, not mistaking them for a washer is very important. I always seem to have several springs that look similar enough that I can't remember which is which, but are different enough that I can't interchange them. In short, when disassembling be sure to catalog and organize parts. I have never regretted doing it, but I have regretted not doing it.

Cylinder Head

Reworking a cylinder head is an important step in any engine rebuild. During the cylinder head rework, the head is first thoroughly disassembled and cleaned. It is then inspected for cracks and any

Engine Removal

One of the most common questions I am asked is: How do I remove the engine from my tractor to rebuild it? For most makes and models of tractors, removing the engine certainly isn't like removing a car engine since the engine block also serves as a structural part of the tractor. In this sequence of photos, Steve Sewell and I will show you how to split a tractor so the engine can be removed and placed on an engine stand for service.

First, you need to round up a bunch of heavy wood, such as 4x4 or 4x6 timbers in 3- and 4-foot lengths, to support the back half of the tractor once the engine is removed. An engine hoist to pick up and suspend heavy components, and an engine stand to place the engine on so you can work on the engine are also needed. After jacking the front end of the tractor slightly into the air and placing the cribbing under it, as seen here, you can begin.

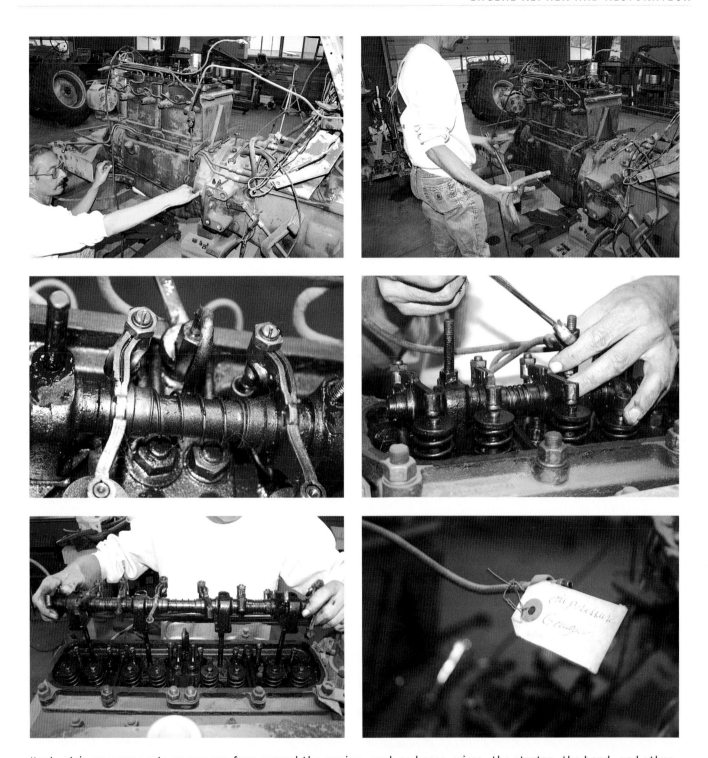

Next, strip as many parts as you can from around the engine, such as hoses, wires, the starter, the hood, and other sheet metal. At least one of the brackets for the engine hoist's cable or chain is typically installed under cylinder head bolts, so you'll need to remove any overhead valve components. Any soft and easily damaged parts, like the oil line that feeds the rocker arm as shown here, should be removed first. Be sure to label your parts. There are simply too many parts and too many that are similar to completely trust your memory.

Engine Removal

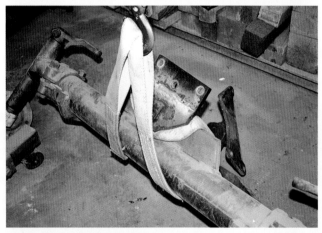

Next to come off the tractor is the front-end. The shop manual for your tractor will need to be consulted for the proper procedures for your tractor. Many front-end components are heavy, and the engine hoist will be helpful in removing those parts. With the front end off the tractor, you can then concentrate on removing the engine.

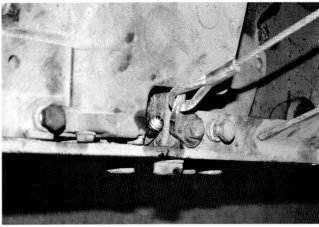

The brackets for your hoist will need to be mounted to the engine, usually under a cylinder head bolt (or sometimes a front or back plate, as seen here) to attach the chains from the engine hoist. You often have to fabricate these yourself. Here Steve is putting the finishing touches on one of the brackets.

Here the bracket is being placed under a cylinder head bolt after a cable stay has been attached. The finished product is shown next.

warpage too excessive to repair. If any of the spark plug holes' threads are stripped, they are repaired at this time. The bottom of the head, the part that mates to the block, is placed on a mill and the bottom is flattened. This helps to ensure that the head gasket will hold the compression gases and last the life of the engine. Another area of the head that may receive machine work is the manifold mating surface. And if performance enhancements are being done at the same time, you should consider having the manifold intake and exhaust openings smoothed and enlarged. This process is known as porting.

Valvetrain

If the engine is a valve-in-head type, the valvetrain is rebuilt at the time the head is reworked. If the engine is a flat head, or L head, engine, the valves are done when the rest of the engine is reworked. During a refurbish, the valvetrain is usually just visually inspected, without disassembly, and only parts that are clearly beyond use are replaced. There usually isn't any thorough cleaning. The seats are lapped if there appears to be any carbon deposits or irregularities on them, and the springs are tested, if possible. Guides are not replaced unless they are known to leak oil. Everything is tightened to proper torque specifications.

Engine Removal

Attach the cable and chains to the engine hoist and raise the engine hoist enough to place slight tension on the cable or chains, but don't try to lift the engine yet. Note that we also have a backup chain in place in case the cable fails.

To rebuild the valvetrain, the individual valvetrain components are all disassembled and cleaned and measured for compliance to minimum specifications. The most important are the valve seats and valve profiles, the valve stem diameter, guide clearance, and valve spring strength. Out-of-spec components are replaced, otherwise they are retained. Past this, the rebuild of the valvetrain is very similar to a refurbishment. During a remanufacture, the springs, valves, and push-rods are replaced as a matter of course without measurement, and the seats usually are replaced. If hardened seats have not been used in the past, you can add this highly recommended item now. To put hardened seats into an engine that did not have them, recesses are cut into the block or head and seats are pressed into place. The rocker arms and rocker shaft are cleaned and inspected. Cam followers are refaced or replaced if they show signs of surface cracking. In all circumstances new keepers are used. If valves are reused, be sure to return each valve to its original location. Drilling holes in a yardstick large enough to accept the valve stems and marked with the words "front" and "back" will keep the valves oriented.

Antique Tractor Tip

Head Gasket Hint
Unless you have evidence to the contrary, always install head gaskets with the lips of the metal edging facing up.

Note on Valves and Unleaded Gas

Many antique tractor engines did not come from the factory with pressed-in valve seats; that is, originally the valves seated directly on the material of the cylinder head or block. If newer nonleaded gasoline is used in these engines and the tractors are then used for heavy work for long hours, the material in which the valve seat can erode, leading to valve seat recession. Extreme cases render the head or block useless. The only cure is to have modern valve seats cut and pressed into the head or block. Modern seats are made of a material capable of withstanding the higher seat temperatures caused by unleaded gas. If you plan on using your antique tractor extensively for hard work, you should seriously

Notice the wood cribbing under the tractor that is supporting the rear of the tractor. It is critical for your safety and the safety of others that you build this cribbing thoroughly and with enough stability and strength to hold the rear of the tractor for the weeks or months it might take you to rebuild the engine. Steve inspects his work and then begins the process of removing the bolts that hold the engine block to the frame of the tractor.

consider having hardened valve seats installed. If your tractor will not be worked strenuously, unleaded gas will not cause any problems for your engine and having newer valves seats installed is not imperative. I recommend installing these seats if at all possible.

Head Installation and Gasket Replacement

Installing the head means obtaining new bolts or studs first. These should be hardened and should be lightly coated with 30-weight oil before assembly. Be sure to wipe away excess oil. If using studs, tighten the studs hand tight and then tighten slightly further. Line up the gasket on the block. The "lip" side of the gasket is almost always placed in the up position (against the head). You should consult your manual to verify this, but lip side up is a safe bet if your manual doesn't cover this subject. The gasket should be installed "dry." That is, it is best not to use a sealant with the gasket. If you insist on using

a sealant, some flat black paint can be used or a thin copper-based gasket sealant. Do not use a thick, high-temperature sealant with the gasket! Line up the head on the gasket (and studs, if your engine uses them) and then tighten the nuts or head bolts finger tight. Following the tightening diagram in your manual, tighten the cylinder head to the torque specifications your manual calls for. If there is no guidance in your manual, tighten the bolts to at least 40 foot-pounds of torque. Also tighten the bolts (or stud nuts) in a criss-crossing pattern that starts with the inside bolts and moves out. Remember to tighten the cylinder head fasteners to torque specifications after you have had the tractor running 15 minutes and again after one hour.

Manifold

There are two primary things to look for in the manifold(s) of an antique tractor engine when you are rebuilding it: You want your manifold to have a flat, defect-free mating

surface (the surface that mates up with the cylinder head or block). You also want to try to find weak areas in the material of the manifold. Checking the mating surface of the manifold is simple. Place the manifold on a flat object, such as a smooth steel table, orienting the manifold so the surface that mates to the cylinder head or block mates to the table. Then go around the edge of the manifold and try to force a very thick piece of paper (I use a matchbook cover) between

Antique Tractor Tip

Cylinder Head Torque
Torque the cylinder head bolts very completely and methodically when assembling the engine. Be sure to re-torque the head bolts after the engine has been started and running for about 15 minutes.

Engine Removal

Once the bolts are removed, you can begin the work of prying the engine away from the frame. This is a delicate step that requires you to pull the engine straight away from the frame since the transmission input shaft seen here in the last photo is, at this point, inside the clutch and still attached to the engine. Preventing the engine from binding against the shaft requires a fairly narrow range of lifting force from the hoist. Once free, immediately inspect the hoist cables and brackets for strain or impending failure.

Once the engine is free, you'll need to remove the clutch from the flywheel. Here the fingers of the pressure place are showing considerable wear and the pressure plate will have to be replaced.

the manifold and the table. You should not be able to push the paper up under the manifold more than 1/8 inch at any place. If you are able to do this, double check the mating surface with a flat edge, like the edge of a steel ruler. If your mating surface is not flat, then check with a machine shop to see if the manifold has warped past the point of usefulness.

Checking for weak areas is easy if you have access to a sandblaster. Very thoroughly sandblast the manifold (being careful to cover the mating surface with duct tape). If heat and rust has eroded any part of the manifold to the point that the walls of the manifold are weak, sandblasting will find and expose this weakness in the form of pinholes. If your manifold survives the sandblasting, probe various areas of the manifold with a ball-peen hammer, gently tapping and listening for the dead thud of weak

Antique Tractor Tip

Block Repair
If your engine block is cracked, have a machinery expert tell you how critical the crack is. Many cracks, especially with sleeved engines, are actually harmless or can be easily and affordably repaired. The same holds true with cylinder heads.

The flywheel is the next part to come off. Most flywheels have some sort of indexing system that forces you to orient the flywheel a certain way when you install it. This is usually accomplished with small offset steel pins or offset bolt holes. However, you should mark the flywheel and the crankshaft with matching marks using a center punch to make sure you can reassemble with the correct orientation if it turns out your flywheel has none of these orientation features. It helps to have handles when you remove the heavy flywheel. To make handles, thread large bolts or large steel rods threaded for the purpose, as Steve has used here, into the pressure plate mounting bolt holes. Have a helper unbolt the flywheel and steady the engine as you remove the flywheel.

areas. If you cannot find any weak areas with these two methods, feel free to use the manifold. By the way, manifolds on Ford 8Ns are notorious for eroding at the very back (closest to the operator), so be sure to very carefully check this area. Pinholes and small cracks can be repaired by most competent welders, but any defects larger than these and you would best be advised to replace the manifold if you can find and afford a replacement.

Block Conditioning

CLEANING

The first step in any type of block work is a thorough cleaning. If you cannot have the block tank cleaned, then the best cleaner for the cylinder walls is (believe it or not) soap and water. Many manufacturers recommend using just soap and water in the cylinders. Simply swab the cylinders liberally with soap and water, rinse

thoroughly, and dry completely. The exterior of the engine can be cleaned with any degreaser of your choosing, and hard-to-reach areas can be rinsed clean with a spray type of degreaser. The real problem occurs, though, if you are going to leave the crank in place. If you are not planning on removing the crankshaft during your rebuild, then you will have a hard time cleaning the inside of the sleeves and cylinders without getting

Engine Removal

The author installs the stand arm to the engine in preparation for placing the engine on the engine stand. Next, Steve places the stand itself on the stand arm. At this point, you are finished and ready to lower the engine and stand to the ground and remove the cables and chains.

cleaner in the main bearings of the crankshaft. Make sure the bearings are taped up with waterproof tape, such as duct tape, or tape plastic around the bearing. Afterward, squirt a water-displacing oil such as WD-40 very thoroughly around all the main bearing surfaces and the bearings themselves.

If you are removing your block from the tractor, I recommend a total disassembly of the engine and then having the block hot tanked. Most full-service automotive machine shops will provide this service if you are going to have them machine the block afterward. It has been my experience that because of the cost and hassle in disposing of the tanking solution, most shops will not clean parts and blocks for a customer who is not also purchasing other services at the same time. Hot tanking involves placing the block for several hours in a tank that is filled with a lye-based solution that is then heated. The only drawback to this is you must be willing to replace all bearings, and so on. This solution will dissolve the soft babbit bearing used for the camshaft, etc. The oil galleries of the block and the crankshaft should be checked for cleanliness afterward. Pipe cleaners, mild air pressure, and pressurized cleaners that readily evaporate, like brake cleaner, can be used to clean them out if needed. Just be sure to prime the galleys with oil afterward, especially if you use a highly effective degreaser such as brake cleaner to rinse them.

Removing Head Bolts and Studs

If your engine uses bolts to mate the head to the block, you should use extra care. Many times they have galled or rusted into place, and no amount of strong-arm tactics will remove them. Avoid at all costs applying so much force that you break them off. Applying heat (and penetrating oil if you can get it to work under the head of the bolt) will sometimes help to loosen these fasteners. Be careful if you plan to reuse the bolts: Cylinder head bolts are heat treated and too much heat will soften them and ruin them for use.

If your engine has head studs, removing the nuts to remove the head often isn't too difficult, but removing the studs themselves can be a problem. Using a special tool called a stud remover is highly recommended as the usual doses of penetrating oil and heat. Use of a stud remover is not easily substituted: I have seen people successfully remove them with vice grips, but I believe the studs are weakened by the process. At any rate, before you jump into doing this, consider whether you really need to remove the studs. If all you are going to do is hone the cylinders and install oversize rings, then removing the studs isn't necessary. Resurfacing the

When you say "combustion chamber," many folks have a mental picture of a fairly large volume inside each cylinder. As this picture of the underside of a cylinder head from a four-cylinder engine shows, the volume is quite small. The tops of the piston will rise to a level nearly even with the bottom of this head before combustion begins in earnest. Each of the combustion chambers you see here consists solely of the dished-out portion plus a negligible volume created by the head gasket. All of the fuel/air mixture the engine brings in to each cylinder during the intake stroke is compressed to this very small space before the spark lights the fuel/air mixture. The difference between the volume brought in and smallest volume after compression is called the compression ratio. Most antique tractor gas engines have around a 9-to-1 compression ratio.

block or performing machining work on the bores requires removing head studs, though. Talk to your machine shop first if you are unsure.

Bore and Sleeve Conditioning

Getting Ready

Reconditioning the bore or sleeves is probably the signature procedure of an engine rebuild. Exactly what steps you take depends on a lot of different circumstances and the type of engine you have. The first step is determining what type of cylinders your engine uses. One type, and the type that becomes more common the younger the tractor engine is, is the bored

cylinder. These cylinders are cut directly into the engine block.

The second main type is called a sleeved engine. These engines have a cylindrical cast-iron "pipe" driven into each cylinder hole in the engine block. The pistons move up and down within these sleeves. There are two different subtypes of sleeved engines, dry sleeve and wet sleeve. The important distinction is that the dry sleeve never touches the coolant of the engine, hence the term "dry sleeve," while the coolant of a wet-sleeve engine does touch the sleeve. Dry-sleeve engines have sleeves pressed into cylinders cut directly from the engine block. This type of sleeve is also used to repair bored block engines. With a wet-sleeve engine, the block is hollow and the sleeve spans the hollow chamber from the top of the block to the bottom of the block. The engine coolant is free to touch the sleeves.

The wet sleeves have special O rings at the bottom of the sleeve to keep coolant from leaking from the block, and at the top of the sleeve is a press fit (very tight and snug) into the top of the block. This keeps coolant from leaking from the top of the sleeve. The sleeves themselves are otherwise much like bored block engines: They are made of similar material and the internal surface has to be honed; oil reaches the cylinder walls the same way. Sleeve walls, however, can only be refurbished a time or two before the entire sleeve must be replaced.

Often performance can be enhanced by using oversized sleeves much in the same way a bored block engine can be overbored. Many times a newer version of your model tractor has a greater rating only because the sleeve walls are slimmer. This allows for greater displacement and horsepower. Be careful not to increase the horsepower of the tractor beyond its design range. This can be very dangerous and increases

the likelihood of rearward flips. Extensive engine modifications should only be done by those committing the tractor to special use, such as tractor pulling, where the danger is mitigated through wheelie bars and controlled circumstances. I do not recommend increasing displacement or horsepower on any antique tractor without careful thought and consideration in addition to receiving training from others experienced in this area.

The extent of the rebuild you're planning for the engine determines the point at which you're ready to rework the cylinders. If you are doing a complete remanufacture of the engine, you need to completely remove and disassemble the block and clean it through the hot tanking method. At this point new freeze plugs would have been installed, but not much else. If you are doing a restoration, the block also has been removed from the tractor and completely disassembled. If you are doing a refurbish, then nothing but the head will have to be removed, and any other components necessary to remove the head. The next step in a refurbish is removing the connecting rods from the crankshafts and pulling the pistons and the connecting rods out from the engine. Disconnecting the connecting rods from the crankshaft must be done from under the engine, but the pistons must be pulled from the top afterward. Breaking the ridge at the top of the cylinder with a ridge reamer will most likely be

necessary first. With some tractors, like the Ford 8N, where the oil pan is a structural part of the tractors and is massive, the work can be difficult. On some engines, it is as simple as removing a small automotive style of oil pan. Either way, gaining access and working safely are the keys. If you are restoring a horizontal John Deere engine, you must remove the connecting rods from the crank from the top access cover.

Initial Measurements

The first task in reconditioning cylinders is simply inspecting them. The cylinders may look glazed over, slightly dirty, and have many light vertical scratches, scuffs, and marks. These are normal and should not be cause for alarm. If, however, your engine has rust that deeply scars the surface of the cylinder walls or there are structural problems such as cracking, gouges from broken pistons and connecting rods, and so on, then you have serious problems that will require new sleeves—or if your engine is a bored engine, the advice and work of a machinist. Some of these faults can be corrected, so do not despair if you find them, but do not put any more time, effort, or money into the engine or tractor until you have determined the extent of the problem.

If your engine checks out visually, you must next take measurements of the cylinders to determine the extent of wear. Unfortunately, accurate, decent-quality micrometers with a proper apparatus for making internal measurements are expensive. Look into renting or borrowing this tool. Only two locations in my medium-sized town rent them. Once you obtain micrometers, the procedure is straightforward: Simply take two measurements halfway down the cylinder at 90-degree angles to each other. Your service manual will tell you the maximum allowable diameter for the cylinder. Just as important, it will tell the maximum difference between the two measurements before you are required to bore the engine. Cylinders do not wear evenly; they wear elliptically. If this elliptical tendency is too pronounced—that is, if the difference between your measurements is too great—then the cylinders must be rebored, even if neither measurement exceeded the maximum allowable diameter.

At this point, between the inspection and the measurements, you have the information necessary to decide what needs to be done to the cylinders of your engine. If you have measurements that exceed maximum, then you must rebore the engine or buy new sleeves. If not, you can simply refurbish the bores or sleeves. If you do this, you may have to buy oversized rings for your pistons. Be sure of the availability of these rings before you commit to honing the cylinder walls. There are several custom piston ring manufacturers across the United States that you can rely on to make the proper rings if your normal parts channel can't get them for you. This is cheaper and easier than reboring or buying new sleeves and should be done if the condition of the cylinders is good enough to warrant it.

REPLACING SLEEVES

If your inspection or measurements indicate the need for replacing sleeves, you actually may have a choice as to how to proceed. Many sleeves will have wall thickness ample enough to allow boring the sleeve. This means you leave the sleeve in place, and then bore out the sleeve to refurbish the cylinder. This may be an attractive option if the sleeves for your engine are particularly expensive or difficult to find, or if you have access to competent, inexpensive machine shop services. An automotive machine shop will be able to tell you if this option is available to you in your particular circumstance.

If boring the sleeve is not an option or would not be a sound decision in your circumstance and you must replace the sleeves, do not despair. Replacing them is not too terribly difficult if you use patience and the right tools. Many sleeve pullers are available on the market, and homemade versions are easy to design and build. This tool works by forcing a metal disc to push up on the cylinder sleeve from below. It is usually attached to the top of the block using any available head stud or stud holes. The sleeves can be difficult to get started, and having patience, using penetrating oil, and periodically applying moderate heat may help. Once started, they usually slide right out. In severe cases, where a connecting rod has broken through the sleeve or some type of deformation or rust has damaged the exterior of the sleeve, the assistance of a machine shop will be necessary.

Antique Tractor Tip

Junk Practice
Use scrap engines to practice valve lapping and cylinder honing skills.

BORING CYLINDERS

If your initial measurements and inspections indicate the need for reboring, then a machine shop can do this for you relatively inexpensively. ONE IMPORTANT NOTE: Make sure you have the proper pistons on hand before you do this. Reboring the engine to .020-inch oversize only to find out that .020 oversized pistons do not exist will frustrate you and waste your money. Another issue to keep in mind with bored engines is the possibility that the cylinders exceed the maximum allowable diameter after reboring. For example, let's say your engine has been rebuilt twice before. It is now bored to .060-inch oversized. Because this engine was stuck and rust had formed on the cylinder walls, or maybe because of excessive wear, you have determined that the engine must be bored. The people at the machine shop tell you that they cannot remove all traces of rust or wear without boring the cylinder past maximum allowable diameter. Now you have a dilemma: You need to rebore but you can't, due to restrictions on the maximum allowable diameter. Does this mean you must throw away the engine block and find a new one? Fortunately, no. You can

This is an engine block to a flathead or "L" head engine. The valves operate within and are seated on (valve seats seen along the bottom) the engine block. This creates an engine with a low, flat cylinder head, hence the name.

a texture to the cylinder wall, improving compression. Honing is a bit of an art, and I recommend practicing on junked single-cylinder engines you get from a local small-engine repair shop. The trick is to match the vertical movement and the proper drill rpm (the honing stones are driven by an ordinary hand-powered drill) so the cross-hatch helical patterns left by the stones are about 30 degrees apart. After a little practice, you should be able to generate the correct pattern of honing marks.

PISTON AND WRIST PIN CONDITIONING

Now that your block and cylinders are ready, you should turn your attention to the pistons. If your pistons are new because you bored or sleeved the cylinders, then you can safely skip this section. If you refurbished the cylinder walls, then you will need to refurbish the pistons as well. Refurbishing the pistons consists mainly of inspecting and cleaning them. To inspect the piston, you will first need to clean it. If the piston is aluminum (which is possible for various reasons), then you must clean it using a solvent such as carburetor cleaner, kerosene, and so on. If the pistons are cast iron (check with a magnet) the best cleaner is a lye-based solution. Don't put aluminum pistons in a lye-based cleaner! Your pistons will dissolve! It may be necessary to use a tool called a piston ring groove cleaner to completely clean the ring grooves. An old piece of piston ring often makes an acceptable substitute for this tool.

have custom-made sleeves pressed into the bore and then have these sleeves bored to the proper dimension. This is a particularly common approach used with engines that have excessive rust formation on the cylinder walls.

Refurbishing the Cylinder Walls

If your measurements indicate the need for a simple refurbish of the cylinder walls, you are in luck. This is a straightforward procedure that most people can accomplish. As always, care, precision, and patience are the watchwords. First, prepare the block for the work. If you are leaving the engine in place and the crankshaft is left in place, be sure to adequately cover it and protect it from cleaners, metal chips, and dust from the cylinder refurbishing process. THIS IS VERY IMPORTANT. Any metal chips or

harsh cleaners that work into the bearings will more than likely ruin them. After this, the next step is to break the ridge at the top of each cylinder. This ridge is left behind because the rings of the piston do not travel all the way to the top, leaving a small area at the very top of the cylinder without wear, resulting in a ridge. Removing this requires a tool called a ridge reamer, and it will make short work of the job. There is no short cut or cheap tool to try here. Fortunately, you will be able to rent a ridge reamer if you don't want to buy one.

The next step is honing the cylinder walls. Honing accomplishes many things: First, it removes surface imperfections and the vertical striations engines develop. Next, it has a tendency to return the cylinder walls to a true circle, minimizing the elliptical wear profile. Last, and perhaps most important, it restores

After soaking and scrubbing, rinse the pistons thoroughly and allow them to dry. Now it is time to inspect and measure them. Check for cracks in the piston at the skirt area and replace any piston that shows a crack or has a noticeable warp to the skirt area. Another general inspection is to check the pin area for cracks and the piston pin keeper clip groove for excessive wear. The most important thing to check for is worn piston ring grooves. This wear will show up as a slant or bevel to the U-shaped groove the ring rides in. Pistons with worn grooves will allow the rings to rock and lean, losing compression and causing premature ring wear and failure. If your pistons show this wear pattern, it is correctable by having a machine shop widen the grooves. The shop will provide spacers to take up the slack the extra width will cause.

The next step is measuring various aspects of the piston to verify that the piston is still usable. One significant area of wear is the piston pin (also called the wrist pin) holes. These holes tend to wear out of round as well as exhibit general wear. You should consult your tractor's service manual for the exact specifications for the pistons in your engine to determine if the wear of the piston pin holes is acceptable. The wrist pin itself should be measured at this time and replaced if worn past specifications or worn out of round. Make sure there are no serious scratches or striations on the piston pin. Now is the time to make sure you have new wrist pin keeper clips to use for assembly.

CONNECTING ROD CONDITIONING

The connecting rods should be reconditioned wherever possible. This means pressing a new piston pin bushing in the rod and using new journal, or connecting rod, bearings. The other important aspect of connecting rod reconditioning is verifying that the rod is true, or perfectly straight, and that the journal end of the rod is perfectly round. Verifying straightness requires a special fixture, and reestablishing the centricity of the journal end requires another special tool. Machine shops have these tools, and if the connecting rods in your engine don't pass your visual inspection, or if the pistons or journal bearings are highly worn, having a shop inspect your rods may be a wise move. I do this whenever I replace bearings, but some folks never do this unless the rod doesn't look right to them. It should go without saying that any rod that has cracks or a noticeable curve or bend must be replaced. Connecting rods are also matched to the cylinder and have an orientation as well. Be sure to note each connecting rod's cylinder and orientation before removal.

BEARINGS, CRANKSHAFT, AND CAMSHAFT
Bearings

In engines, crankshaft bearings in the inner parts of the engine are not of the ball or roller type (with a few very unusual exceptions), but instead are a babbit (lead alloy) based material that, with the presence of oil, is incredibly smooth and long-lived in extreme conditions. The early bearings were simply poured directly into the cavity of the bearing caps and mounts. Pouring babbit into these older style engines requires the services of someone trained in pouring babbit, and you should consult local automotive machine shops. Often these shops know of folks in the area who do this type of work. In most of the younger antique tractors the bearings, while made of the same material, are made in layers, with exterior layers made of a harder alloy. This allows the bearing to retain its shape and be handled as a separate part. This type of bearing is commonly called an insert bearing.

To renew crankshaft bearings, first remove them and measure their thickness and check for unusual wear. The service manual for your tractor will have the thickness specifications. It is possible to check the bearings and crankshaft with the crankshaft in place on many tractor engines. Simply check the bearings one at a time, replacing each bearing before moving on to the next. Removing the half behind the crankshaft can be accomplished through the use of a bearing pin, which is simply a small wedge-shaped pin placed in the oil gallery hole of the crankshaft. Rotate the crankshaft to remove the bearing. With the bearing out, measure the crankshaft and compare its wear and out-of-round measurements to the specification. Have the crankshaft reground if any of the main or rod journals do not pass. New bearings will be required if the crankshaft is reground. While measuring the crankshaft, check for any striations, scratches, or any unusual galling or corrosion. Even if the crankshaft does not need to be reground, you may want to have it

When the word "crankshaft" or "camshaft" is mentioned, now you know what each one looks like. Above is a crankshaft from a Farmall tractor's engine. Below is a camshaft from the same engine. The offsets seen on the crankshaft are what the pistons drive against via connecting rods, and the lobes along the camshaft are what open and close the valves.

polished if any of these conditions occur. All the bearings of the crankshaft, including the journal bearings, are checked this way.

Camshaft bearings are a bit different. These bearings, while made of the same material, must be pressed into place. In addition, many replacement bearings for these engines must be reamed. Reaming is done to establish final clearance of the bearing material after it is pressed into place. This process removes material from the bearing to reestablish this clearance and brings the bearing back into round. Checking to see if they need replacing is tough. Even engines that originally did not have the precision style bearings from the factory often have replacement bearings available that are of the precision type. In short, reaming is not always necessary when installing camshaft bearings. If reaming is

necessary, the procedure is best left to a machine shop. One other note: check the wear of the camshaft journals. Oversized camshaft bearings are often unavailable, necessitating the replacement of the camshaft if the journals have worn beyond specifications.

Camshaft and Valve Assemblies

The next area of inspection and renewal is the valve activation and lifter system. The primary duty of the camshaft is to operate the valves, though it usually powers other systems, too, such as the distributor, governor, and oil pump. Whether your tractor has an overhead valve engine or a "flathead," or valve-in-block engine, you will have a lifter system that operates the valves in the engine. This system always begins with camshaft followers (most often called tappets in antique tractor

literature, though followers seems to be in more common modern use) that ride on the camshaft lobes (the raised portions of the camshaft). These tappets operate the valves or operate assemblies, which in turn operate the valves.

A tappet in a flathead engine is simply a small steel cylinder with an adjustable striker that hits the valve stem. The bottom of the tappet, the portion that rides on the camshaft, may need to be replaced or renewed. Renewing the camshaft face of the tappet is called refacing. Refacing tappets is necessary if the metal has taken on a fatigued or discolored appearance. Any sign of cracking or stress fracturing is reason to discard the tappet. If the bottom of the tappet looks to be in reasonable condition, and the adjuster works well, then the tappet may be reused. In a complete remanufacture, tappets are always refaced. Each tappet should be returned to the camshaft lobe it came from.

In valve-in-head engines, you must also inspect the pushrods. These rods ride on the tappets and in turn move a rocker arm that operates the valve. Inspect both bearing surfaces on the pushrods and also inspect them for straightness. The bearing surfaces are checked just like tappets: If the surfaces are free of defects and appear to be a bright metal color or a very light gray, they are fine. To check for straightness, simply rotate each pushrod between your palms very rapidly. Your hands will be able to detect the slightest bend better than your eyes. If available, a machinist's V block and a dial indicator will give a very accurate assessment of straightness.

The rocker arms must move smoothly and without play. Check the spacers, shims, and any springs within the rocker shaft assembly for wear and weaknesses. In horizontal John Deere engines, check the push-rod tubes for any signs of fatigue or failure. Failure of the tubes will allow coolant to enter the engine oil reservoir and combustion area.

Rocker Arm/Lash Adjustment

Lash is the space that exists between the rocker arm and the valve stem, or the tappet adjuster and the valve stem in flathead engines. The most important consideration in this adjustment is determining whether the valve lash should be adjusted while the engine is hot or cold. The proper approach can vary across models from the same make. Consulting a service manual is the only way to know for sure what lash procedures and specifications must be followed. Most manufacturers called for the lash adjustments to be made when the engine was hot, so if you can't find directions otherwise, hot is best. In the absence of specifications, I would use .010 inches for the intake valves and .16 inches for the exhaust valves. An engine should run with these specs without causing harm. Use this to hold you over until you can find a manual with the manufacturer's specifications. The remaining steps for setting the lash can be found in the maintenance section.

TIMING GEARS
Front Cover

Within every engine lies the crankshaft and camshaft gearing. In modern engines a chain or a belt may drive the two, but in every

antique engine I can think of the camshaft gear is driven directly by the crankshaft gear. The cover on these gears is called the front cover, and gears on the shafts are called timing gears. For this reason, the cover is also called the timing-gear cover. There are several issues with the front cover and timing gears to be aware of as you restore your engine. The gears must be synchronized before the system will run. Synchronizing the gears involves lining up marks as you reassemble so that the valves synchronize with the pistons. Your service manual can further describe the marks. The most important thing to realize is that you may have to make the marks yourself before you disassemble. If that is the case, a sharp, cold punch will make adequate marks for you.

There are a few other things to keep in mind when rebuilding the front cover assembly. The moving parts behind the front cover are lubricated by a device called an oil slinger, a saucer-shaped washer installed between the front cover and the block. The slinger must be oriented correctly for this to take place. Be sure to consult your

manual for details. A seal on the front cover prevents oil from leaking past the shaft. Carefully press this seal into place, making sure the seal is straight and the lip faced toward the crankshaft gear. Over the years this seal often wears a groove in the crankshaft, allowing even a new seal to leak. Repairing this requires a thin metal tube called a sleeve or special repair compounds that can be applied and machined when the crankshaft is out of the engine and at a machine shop. Often there is a dust seal made of felt that is installed after the seal. Last, install the pulley as tight as possible, but do not use excessive force. An excellent choice is a pneumatic air impact wrench. Use about 150 foot-pounds of force, unless your owner's manual provides a different setting.

SEALS

The oil seals for the engine should be replaced during every rebuild. The rear main oil seal for most antique tractor engines will be a two-part seal that is a tar-impregnated rope (traditionally jute). This is often called jute packing. The packing can be fed up and over the crankshaft

Here are the rod and piston assemblies for a late 1940s model John Deere. Since these engines only had two cylinders instead of a more traditional four or six, the needed displacement came from much larger bores and slightly longer strokes than traditional engines with the same horsepower rating.

if the crankshaft remains in place (there are tools to assist with this). Some engines have removable seal carriers. They can be driven out, the seal replaced, and then the carrier rotated back into place. The bottom half of the seal is placed in the bottom seal carrier and the bottom carrier is simply bolted into place. Jute-packing seals must be flattened before they are installed into their carriers, and then slightly crowned after they are placed in the grooves. To aid in preventing an oil leak where the two jute-packing strips meet, creating two seams, the bottom carrier jute packing is trimming slightly long, approximately 1/4 inch.

Oil Pump

Oil pumps in antique tractor engines are almost exclusively the gear style pumps. These pumps use two gears driven in unison to create an area of high pressure on one side of the gears and low pressure on the other side. Therefore, oil-pump rebuilds should focus on making sure the gears and their drive system are in good shape. This means making sure the gears are not worn, the bushing for the gear drive shafts is not worn, and the pump housing passes a visual inspection. First clean or replace the pickup screen. Disassemble the pump housing's bottom cover, which is often the pickup screen itself. Note the gasket thickness for later reference. Save any parts of the gasket that are salvageable. The gears should now be revealed. Measure the clearance between the tips of the gears and pump housing. There should be a maximum clearance between these two or else gear set replacement is necessary.

Next, check to make sure the pump housing is solid and not cracked. Hold the driven end, the end that rides on the camshaft, and move it back and forth sideways while the gear housing is in a vise. There should be only very slight movement side to side. After you have determined that the pump is in fine health, then reassemble, being sure of one thing: The gasket that was between the pickup screen and gear housing was not only a gasket, it was a spacer. This gasket, if too thick, will cause very low oil pressure and may even result in engine damage. Be sure you use a new gasket that is very close to the same dimensions as the original gasket. Remount the oil pump before mounting the oil pan.

Oil Pan

Oil pans are notorious for leaking, and replacing the oil pan gasket is important to a leak-free engine rebuild. The problem with most oil pans is they have weak sides and flanges that warp and bend over the years. These pans must also seal around the curves of the seal carrier blocks and other structures. In addition to the primary oil pan gasket, there are often secondary seals or gaskets placed where the pan meets the curved portion of the engine seal carrier blocks. To prevent leaks, all of these gaskets should be new and supple, mating edges should be straight and clean, and gasket sealant must be used. Many people use the gasket-forming products now available for their superior ability to prevent leaks. If originality in your tractor restoration is important, you shouldn't use these products; otherwise, they are a good idea. Never use these products as a substitute for large or thick seals. These are adequate in replacing flat gaskets only.

FLYWHEELS
Truing, Installing

The flywheel at the rear of the engine, if used for clutches, should be resurfaced. John Deere two-cylinder engines are the exception, as their flywheels are not used for the clutches. In addition to resurfacing, the flywheel should be rebalanced if possible. If your flywheel is out of balance by more than a little bit, balancing it will increase main seal longevity, main bearing longevity, and dramatically reduce the vibration your engine produces. These chores can be handled at any automotive shop. The pilot bushing for the driveshaft is often in the flywheel itself. Pulling the bushing can be done with a pilot bushing puller. One trick I have seen is to install a grease fitting in the bushing, and then you can use the hydraulic pressure of a grease gun to force the bushing out. Renewing the bushing will simply entail pressing a new one back into place. If the engine timing marks are on the flywheel, then it is very important that you install the flywheel properly so as to synchronize the marks with the crankshaft. As with the crankshaft and camshaft gearing, you often have to create register marks on the crankshaft flange and flywheel before you disassemble.

Startup Procedure

Once the engine is restored, along with its fuel and electrical systems (see next chapter), you are ready to start it. There are a few things you should do to get it ready. The first is called oil priming. This delivers oil to all areas of the engine. While many elaborate schemes can be developed to spin the oil pump to get oil everywhere, simply using the starter to turn the engine until oil pressure is established is an easy, safe way to do it. This, of course, forces you to have a working oil pressure gauge. Once the engine has built up oil pressure, you are free to try to start your engine! This is the most exhilarating or most exasperating activity, depending on whether the tractor starts or not. Make sure the radiator cap is off when you do this.

Once the engine starts, watch the oil pressure gauge carefully, and be sure the engine is maintaining pressure. Move the throttle back and forth to be sure of responsiveness, and check the exhaust for any signs of smoke. Excessive smoke that lasts for more than a few moments is bad. After all seems well, especially at the oil pressure gauge, check the radiator. You should see no active, persistent bubble formation. If all is still well, run the engine for 15 minutes. During this time, do not make too many changes to the engine, though some simple adjustments such as the idle-speed can be made. Listen to the engine carefully for any sign of problems and check for oil leaks. Continue to check the radiator for any signs of bubbles after the engine has warmed up. These bubbles could

Rear main seals of most engines have their own housing or carrier and can be removed separately. This engine has a single piece carrier that holds a seal, while many have a two-piece carrier that houses two pieces of a tarred rope-type seal.

indicate a blown or loose head gasket. Also check for any drop in level. There will be some bubbles and a drop in the level of the coolant as the coolant finds its way to all the nooks and crannies of the engine, but the bubbles should stop within a few minutes and the level should not drop much. Pull the oil dipstick, making sure coolant isn't going to the engine oil. Immediately stop the engine if you see any of this behavior.

After 15 minutes, shut it down, and let the engine cool down completely. Tighten the head bolts or stud nuts back to factory torque specifications. Check the tension of any drive belts and check again for any sign of oil in the coolant or coolant in the oil. If all is well, restart the engine and listen again for any sign of trouble. If the tractor is operable, I suggest beginning the break-in process. If not, rest the engine until you address any problems you have identified. Once you have the engine in good running form you can concentrate on repairing or restoring other

Water pumps have changed little over the years, although some of the shafts and bearings are getting tough to come by. The last pump I rebuilt required a new shaft and bearing assembly. The proper one was not available so a bearing assembly from another application was used, but one side of the shaft had to be cut by the shop before it could work for my engine. The cast-iron claw you see here on the pulley is a stand-off for the fan blade.

aspects of the tractor. During the first few hours of operation, the piston rings have not seated against the cylinder walls. This means that the rings will be letting some oil by into the combustion chamber and some combustion gases through to the crankcase. Therefore, you may see some smoke from the crankcase breather or the exhaust pipe. This smoke should be very light and should disappear within the first few hours of operation. If not, your rings are not seating properly or oil is getting into the combustion chamber some other way. After the first hour, change the oil, and change the oil frequently for the first 100 hours of operation. After that, the engine oil can be changed as usual.

Break-in

Engine break-in has two distinct phases. The first is a short-term period during which the components within the engine wear and settle into place and initial problems reveal themselves. The second phase is longer term and simply establishes the wear pattern the engine will generate and continue to follow through the rest of its life. In the first phase, new piston rings seat, the head gasket expands, and the head bolts stretch, among other things. Piston ring seating is especially important. Chrome-plated piston rings are difficult to seat. In fact, your new engine may burn oil for the first few hours until the piston rings seat.

After the first 15-minute run, finish assembling the tractor so it can be broken in. You just need to complete enough of the tractor to take it out for road tests. Restart the engine and alternate throttle speeds until the engine warms up. After that, operate the tractor in a varied set of road circumstances until the engine has about an hour on it. I recommend occasionally lugging the engine during this time. Do this by moving down a smooth area in road gear and applying the brakes for a few seconds. I do this a couple dozen times during the first hour This will open up the throttle completely and generate the stress necessary to seat the piston rings. After the first hour, tighten the head bolts again; adjust as usual, waiting for the engine to cool before doing so.

Over the next few hours, the engine should not be used for excessively difficult work. After that, the engine can and should be occasionally used for hard work that forces it to lug or brings output power to the fullest of its capabilities. This should not be continuous, though, and should be done in short, sporadic bursts. After 10 hours, the tractor can be used normally. Antique tractor collectors often do not break in their tractor engines properly because they usually have no difficult work for the tractor. Even collectors should lug their engines during the critical first 10-hour break-in period.

CHAPTER 14
Fuel and Electrical System Overhauls

Restoring your engine's mechanical components only brings it part of the way back into service. To run and run well, it will need fuel and electricity at the right times, in the right places, and in the right amounts. Accumulated deposits, sediment, water, or improper adjustment can reduce fuel delivery or interrupt it altogether. Likewise, problems developing or delivering spark, or sparking at the wrong time, can cause your engine to misfire or quit. At their worst, problems with fuel delivery or your engine's electrical system can be hard or even hazardous on your engine. Let's look at the major components of both of these systems and some procedures for restoring them to optimum performance.

Governor Overhaul

The governor is usually just a mechanism that takes the inertial force of a rotating mass, applies it against a spring, and based on the tension of the spring, adjusts the carburetor throttle shaft to reflect the engine speed indicated by the throttle position the operator has selected. The mass is usually

Notice this carburetor is smaller and the fuel line is not your typical 1/4- or 3/8-inch steel fuel line. This is an LP gas carburetor or mixer, as it is occasionally called. LP engines operate on the same principles and dynamics as their gasoline counterparts.

When an engine with a magneto starts, the speed of the engine is slow enough that the armature in the magneto is rotated too slowly to generate a strong, healthy spark. The answer to this is called an impulse coupling. These work by using springs and pawls to momentarily hold the magneto back as the engine starts to turn. Then the impulse coupling trips and drives the armature quickly when the firing cylinder is right at top dead center. The lag angle mentioned elsewhere determines exactly how this accelerated movement is timed to the engine.

steel balls or cams with weights at the ends. The governor is almost always driven by the camshaft. Restoration of the fuel system includes inspecting and replacing, if necessary, the drive gear, the bearings, and the seals, and any part of the linkage to the carburetor that is worn or bent. There are usually two screws on a governor as well. One is called the bump screw. This is the screw that the governor will bump against when it reacts to differences in load for the tractor. This screw is adjustable and your manuals will illustrate how or when you should adjust it. There is often a second screw that most manufacturers recommend leaving alone. This is the limiting screw and it changes the way the governor reacts. This screw can be very difficult to get back into adjustment if you change it. Check your manual for your tractor to determine what adjusting screws you have and what adjustments need to be made. Consider replacing the throttle-to-governor spring at this time. Replace it only if an identical spring can be found. If not, try to reuse the same spring. If the spring must be replaced, and an identical replacement cannot be found, a slightly more tense spring should be used instead. Never use a less tense spring as this will cause your governor to "hunt."

Fuel System Overhaul

FUEL TANK RESTORATION

The best first step in rebuilding a carburetor does not even involve the carburetor! Most carburetor problems are related to the buildup of dirt, grime, varnish, rust particles, scale, and so on that originate from the gas tank and gasoline. Before rebuilding the carburetor, you should first clean the gas tank if it needs it, and replace the fuel line if necessary (and it often is necessary since water in the fuel will have a tendency to accumulate in the fuel line if the tractor sits, causing rust in the lines). If your gas tank has rust in the interior, you should remove the rust and seal the interior of the tank.

Sealing a gas tank involves three steps. The first step, scouring, involves agitating the tank, first filling it one-third full with any lye-based cleaner, plus water, smooth creek pebbles, or roofing nails. This will scour away 99 percent of the rust and remove any gas-based residues. The cleaner is only needed for the first 15 minutes or so. Rinse out the tank and pebbles with water, then fill the tank again with just clear water. Continue rinsing and refilling with clear water every so often. Scouring is tedious and time consuming and takes hours to remove most of the rust. Finding an automated way to do it will help tremendously. I have seen a tumbling apparatus that rotates the tank slowly, much like a rock tumbler. Carrying the tank on a riding lawnmower while mowing, or carrying it on another tractor as you work the fields, will help to agitate the tank. You can stop when the water rinses clear and no large rust flakes are coming out.

After scouring the inside of the tank, you may find that your tank has rusted through at the bottom and leaks. Before, the varnish, grime, and rust flakes were keeping it from leaking. Sealing the tank will close most small pinholes. If the holes are larger than a BB, or you have many concentrated in a small area, you may want to take the tank to a welder and have a repair patch welded in. Because of the cleaning process, and because tank welders have equipment and procedures to lessen the risk, welding the tank will be safe. However,

NEVER WELD A TANK THAT HAS NOT BEEN THOROUGHLY CLEANED! AND NEVER WELD A TANK YOURSELF WITHOUT PROPER TRAINING!

After the scouring step, you will need to etch the tank with acid to stabilize any remaining rust. Use phosphoric acid or any acid metal prep solution from an auto body store for this. It is very important that you very safely and carefully slosh the acid with the gas cap removed. The reaction of the acid to metal creates a gas that can cause the tank to explode or acid to spray out of the tank outlet or spray out from the cap vent hole(s) if the resulting pressure is not allowed to escape quickly. Make sure the acid touches every part of the interior of the tank. Rinse completely several times with clear water and discard the solution properly. Dry the interior of the tank quickly to prevent the formation of flash rust inside the tank. Since it is very important that the interior of the tank is dry for the next step, you should use some high-volume warm-air source for drying. I usually have the best luck with the exhaust from vacuum cleaners, though others use hair dryers.

When you are positive the interior of the tank is completely dry, apply the sealer. This simply involves pouring in enough to adequately cover the inside of the tank and then turning it upside down to drain. These products usually require several days of cure time before gas can be put in the tank. While sealing a tank like this is tedious, time-consuming work, you should do it if you are serious about restoring the tractor properly. A good short-term fix for a rusty, grimy gas tank is to adapt a modern fuel filter to the fuel line. In the long run, the sealing process is much more dependable and will protect your gas tank for a long time to come. To finish the fuel delivery restoration, thoroughly clean the sediment bowl assembly and replace the gaskets and screen. Finally, using solid steel line (never use copper line!), fashion a replacement gas line. Usually 5/16-inch or 3/8-inch line is sufficient. Match the line on the tractor if you are not sure. If you must custom-make a new line, you'll need a tubing flaring kit capable of performing double flares.

CARBURETOR

Start the carburetor restoration by choosing a clean, well-lit work area that allows you to carefully disassemble the carburetor. Then turn off the fuel flow from the tank. Disconnect the fuel line. Disconnect the governor linkages and any choke control cables. Remove the carburetor from the manifold and air cleaner, keeping the carburetor vertical so gas inside the carburetor does not spill. Carefully place the carburetor on the work bench, and remove its top half. Now take the gas that is still in the bottom half and return it to a gas tank. Carefully disassemble all pieces of the carburetor, keeping them in order and making notes as you go. The most important thing at this point is being sure to disassemble as much as possible. This includes the seats, needle valve, idle-adjustment screws, floats, and so on. When removing the mixture screws, turn them down until they seat, noting how many turns it took. This will provide a starting adjustment upon assembly if one isn't available in the manuals. Do not remove the small brass

Here are shots of the drive systems of magnetos. The large U-shaped magnet is called the field magnet, and the components seen here as part of the shaft tell us the shaft is an armature.

On newer magnetos, the field magnets are laminated, as seen here on each side of the case. The bronze-looking tube is the bushing for the magneto driveshaft.

Here is a look at the inside of a magneto in the area of the points and condenser. As you can tell, it isn't much different than a distributor. A lobed shaft opens and closes the points as the engine rotates the shaft. When the points are closed, the coil(s) (not installed here) charge. When the points open, they discharge their current to the spark plug. The condenser, also not pictured, is installed here too.

The lag angle of magnetos is set by rotating the housing relative to the drive assembly. The lag angle is important when the engine is starting and is different for most applications. Graduated marks and a pointer on the drive assembly indicate the angle in this photo.

pieces in the passageways. These are called jets, and at this time, it is best to leave them in place. Soak the components in carburetor cleaner overnight. After cleaning, sort the parts according to your notes and let them air dry.

Now we will attempt to clean out the passageways in the carburetor. The small brass jets that you can see down in the passageways serve as gates for these passageways, and the size of their opening is critical. Removing them can be very tricky, as they are soft and easy to deform if you use the wrong size of screwdriver. I will pass on a trick I learned from a friend: Use a cheap set of screwdrivers to remove the jets. This way, you can use a bench grinder to grind the screwdriver to the exact profile needed to remove the jet without stripping or deforming it. Removing the jets makes cleaning the passageways much easier, though you can often clean them without removing the jets. Some jets, by the way, are pressed into place rather than screwed in. It is impossible to remove these without ruining them, so you should leave

those types in place unless you are sure they need replacing. Heat is also often required to remove the jet. Gentle heat such as a plumber torch is all you would want to use.

The carburetor body parts are usually made of soft metal that is easy to scar and scratch. For that reason, use soft materials, such as monofilament fishing line (30- to 50-pound test) or compressed air to clean out the passageways. Never use hard wires such as steel wires. Compressed air works well, but you can create serious problems if you direct the air into the passageway from the wrong direction. For example, if there is dirt in the passageway, and you do not blow it out from the direction opposite to the normal gas flow, you may create a serious obstruction by wedging the dirt farther up into the passageway. Since the variation in carburetors is extensive, I can't give you a good rule of thumb regarding direction, but gas originates from the bowl, so blowing air toward the bowl is best. There are innumerable methods for dislodging dirt and grime. For example, I have whittled down a plastic cable wrap before and it

After finishing up electrical connections, clamp the condenser down with its retaining bracket and make sure all the wires are down and out of the way of the shaft and are clear of the edges of the cover. We don't want them to get pinched when the cover is installed. Then install the rotor, which is a small plastic cap that sits on the driveshaft and distributes the spark to the correct cylinder. The rotor can only go one way on all distributors and magnetos to ensure we don't alter engine timing by incorrectly installing the rotor.

has worked. Ingenuity is the key. Eventually, a method will work if you take your time and work carefully.

The first step when reassembling the carburetor is making sure the mating surface between the two halves is straight and true. Using a large mill file, you can flatten the mating surfaces. Be careful to remove all particles from the carburetor that result from the filing. Reassemble your carburetor according to your disassembly notes. You should use a rebuild kit to reassemble your carburetor. These kits will, at a minimum, include things such as a new float needle valve and seat, throttle bushings, and so on. The important adjustments to make during rebuild are to the float height and idle needle. The float height specification is usually listed in your tractor manufacturer's service manual. Lacking this, you can use the specification that came with the kit. This specification usually differs from the service manual, so use it only as a last resort. The float height can be set using the small ruler that comes with just about every rebuild kit. Adjusting the float usually means bending the float arm until the proper height is reached.

Other things to make sure of during assembly are that the throttle is properly placed (it's sometimes possible to get it backward) and the springs and bushings were replaced and installed correctly. When you are putting the two main halves back together, make sure the float does not rub against the bowl of the carburetor. You should be able to turn

the carburetor upside down and back several times and hear the float move freely up and down. The idle needle is next; it usually has a set number of turns that will allow the tractor to run, and then once running, finer adjustments are made. This number of turns is found in the tractor's service manual and usually not in the instructions with the rebuild kit. If neither the rebuild kit nor your manual specifies this, use the number of turns you noted at disassembly.

Remount the carburetor and reattach the fuel lines, choke control, and governor rod. Open the fuel flow, watching first for a stuck float that would allow the carburetor to flood with gas and cause a fuel spill. If you are satisfied the carburetor isn't leaking, you can test the tractor by trying to start it. If it doesn't start, and you are sure the problem is fuel related, shut off the gas, remove the carburetor, and open it up. If gas is not present, the float is sticking closed, probably because it's rubbing against the side of the bowl. Readjust the float, making sure it is straight and will move freely in the bowl. If there is gas in the bowl, be sure all jets, orifices, and screens are in place and properly installed. If that isn't the problem, you probably need to reclean the passageways of the carburetor. Once the tractor is running, you can then follow your service manual's procedures for adjusting the carburetor.

Diesel Power Issues

While many diesel engine components and systems are the same as those used on gas engines,

many are different. Diesel engines use compression to burn fuel, and there are no systems on the engine to create a spark. Shutting off the engine therefore requires shutting off fuel flow to the engine. Gas engines deliver the fuel to

Coils are installed using two or more hold-downs. These should be snug and hold the coils securely since the coating on the coil and the coil itself can be damaged or have its life shortened through excessive vibration.

Here is a coil used inside a magneto. It is not functionally different from external larger coils you may have seen in distributor ignitions, although they obviously differ in shape, size, and materials.

The magneto cover and gasket are the last to be installed. Be sure to use care when tightening the cover bolts as you can crack the cover using too much force.

the cylinders via a carburetor, whereas diesel engines require a fuel-injection system to deliver fuel to each cylinder. Carburetors were never used with diesel engines because diesel fuel is a heavier fuel with poor vaporization characteristics. This means a carburetor will not mix diesel fuel and air very well. The fuel-injection system of diesel engines is their Achilles heel, and repairing it is very expensive.

There are many high-precision parts needed to pump and meter the fuel and inject it into the cylinders. These parts receive their lubrication from lubricating additives in the fuel, and care has to be taken to use fuel with sulfur or other chemicals that will help to lubricate. Another concern

with diesel engines is the risk of trash and water in the fuel. Trash in the fuel may score and ruin expensive injection pumps and metering mechanisms. Likewise, water that sits in the fuel system can cause rust to form on these components. Because of this risk, you will never see a diesel engine without extensive filtering systems to remove these impurities.

Rebuilding these systems and components should only be done by qualified shops and trained personnel when a problem actually arises. While many activities, such as replacing injectors, O rings, and so on, can be performed by most owners, these are more maintenance activities rather than rebuilding activities. If the system was working when you started the engine rebuild, you should simply perform any maintenance on the fuel delivery system required by the owner's manual and then wait to see how the fuel-injection system performs after the rebuild. Any smoking or hard starting may be repaired by the rebuild.

Another common device on a diesel engine is a turbocharger or blower. This device helps pack more air in the cylinder, increasing performance. Driven by exhaust gases, this device spins very fast and runs very hot and is very sensitive to a lack of lubrication. When rebuilding the engine, be sure you have thoroughly cleaned the turbine area and have made sure the oil delivery lines or galleys are clear and ready to operate. Be sure the cylinder head gasket is new and the cylinder head is properly installed and tightened

when rebuilding a diesel. The compression ratios of diesel engines are much higher than gas engines, and blowing a head gasket is a very real possibility if the head bolts or stud nuts are not properly tightened and issues such as head warp are not addressed during the machining phase of the rebuild. The last item is the glow plugs. These devices, located in the intake manifold, warm the incoming air and are used when the tractor is being started. They are often burned out in diesel engines and require replacement.

Magneto Overhaul

Overhauling the magneto should be done at every engine rebuild unless the magneto has already been recently rebuilt. There are many subsystems and parts you can renew and replace that will put new life into your magneto. This type of work would be considered a refurbish, though, and if your restoration or magneto require a complete remanufacture or restoration, you would be wise to consider professional assistance. A complete rebuild or remanufacture of a magneto requires a certain technical aptitude, patience, and special tools. The magneto is probably one of the more delicate parts of an antique tractor (though definitely robust compared to most of today's modern electronics!) and many of the subassemblies are difficult to remove and service. To make matters worse, the magneto is probably one of the top ten most expensive parts on your tractor. While I am not trying to discourage anyone, a disassembly and remanufacture of a magneto is careful, deliberate work that should not be rushed and can easily be botched by impatience and poor workmanship.

To refurbish your magneto, you should consider replacing the oil seal, coil, and impulse coupling spring. The points, condenser, distributor cap, rotor, and wires should all be replaced as well. The coil can be replaced and sometimes is fairly easy to access. It can be soldered into place easily by anyone with soldering skills. Unfortunately, it often is difficult to access and may require professional removal and installation. The impulse coupling should be cleaned and oiled. The coupling is also often difficult to remove, but frequently can be thoroughly cleaned with a spray cleaner and lightly oiled while it is in place. This would be the time to sleeve any shafts in which an oil seal has worn a groove and to test the bearings and bushings to consider their replacement. In short, careful inspection, cleaning, and renewal of selected parts can go a long way toward putting zip back into your magneto.

Installation and Timing

Timing magnetos to the engine can be a bit involved, but it is not overly difficult. Since the engine isn't running, you will have to time the magneto with a procedure called static timing. Static timing involves manually creating a spark when the number 1 cylinder has been positioned at top dead center. The magneto is then installed so the unit is properly oriented and the distributor rotor points directly to the terminal for the number 1 spark plug wire. At this point, the magneto is indexed to the engine, but still isn't timed. To time the magneto, you usually rotate the engine until the next piston in the firing order has reached the top of its compression stroke (top dead center), and then turn the magneto body in the same direction the rotor turns, until the impulse coupling fires. This will make an audible click, so it isn't necessary to actually test for a spark. Just listen for the click. While turning the engine to the next compression stroke seems counter-intuitive, this step winds up the impulse. The step wouldn't be necessary if there were some way to wind the impulse coupling before installing the magneto.

RESTORING A DISTRIBUTOR

The rebuilding of a distributor is not as difficult as that of a magneto. Distributors usually consist of just a shaft with a drive gear on one end and a weight and a cam on the other.

The magnets inside a magneto lose strength over time, especially when the tractor sits for long periods of time. The types of magnets in these older magnetos can be recharged. A device designed to do just that is shown here.

This all passes through housing bushings. On a breaker plate, or cam plate, there is a point set and a condenser for controlling the spark. There is also usually some type of dust cover for the cam plate and often some type of oil wick at the shaft hole in this plate for applying small amounts of oil. Removal of all of these parts is fairly self-explanatory. To begin remanufacture, inspect and clean all parts. Plan on replacing the points, condenser, and any washers and nuts. The cap, rotor, and spark plug wires will also

Antique Tractor Tip

Magneto Wisdom
For a more detailed look at the lost art of magneto repair, check out Neil Yerigan's book, *How to Restore Tractor Magnetos*. This is arcane stuff, and Neil is one of the few guys out there who knows these things inside and out.

  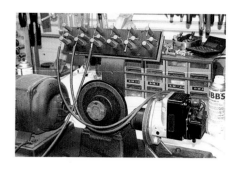

Weaver's Magneto Repair uses a magneto test bench to test their rebuilt magnetos or to troubleshoot newly received magnetos. There are six spark rails available for up to six cylinder magnetos and are set to match the spark gaps testing specification call for that particular magneto, although there isn't necessarily a gap that matches any particular tractor. The test bench spins up the magneto, and as you can see here, this magneto is throwing a nice, healthy-looking blue spark. This is what you see when you are in your shop testing your tractor's ignition with a spark plug.

be replaced. After cleaning, remove the drive gear from the bottom of the shaft. Removing the drive gear usually involves pressing out a pin that holds the gear to the shaft. Once done, the whole shaft will slip out. The bushings should be replaced if there is any play in the distributor drive shaft. There are usually two bushings, one at the top and one at the bottom. The one at the top is usually worn, but the one at the bottom usually isn't nearly so worn and may be fine. Replace any springs on the flyweights that are broken or are clearly weak. These springs must be replaced by identical springs, and they must be replaced as sets.

To static time a battery-coil ignition system, you basically follow the same steps as static timing a magneto. You first position the engine, then you index the distributor to the engine, and then you time the distributor. The only differences are that there is no impulse coupling to wind, and setting the position of the distributor can be done with the engine in a single position. The next difference is that the index

point of the distributor is with the engine at the idle timing mark of the flywheel, and not at top dead center. To begin, remove spark plug number one and rotate the engine until cylinder number one is generating compression. (I place a light piece of paper over the spark plug hole. It will lift when the cylinder starts generating compression.) Continue turning the engine until the idle timing mark of the flywheel (usually about 2 to 8 degrees before top dead center) lines up with the timing reference pin or mark. Turn the drive shaft of the distributor until the rotor points to the number one terminal of the number one spark plug wire. At this point, the mechanism that turns the distributor shaft should have recesses or lugs that line up the drive shaft of the distributor. Mount the distributor and leave the mounting bolts finger tight. Reassemble the distributor cap and wires, and ground the spark plug of cylinder number one. Turn on the ignition, and then rotate the distributor opposite the direction of rotation of the rotor. Do this

slowly until you see the spark jump the gap at the spark plug. Tighten down the distributor at this point. Finish the timing by performing a running timing adjustment.

GENERATOR AND STARTER OVERHAUL

From a financial point of view, unless the only thing your generator needs is a set of brushes and bearings, rebuilding the generator/alternator and starter motors is best left to the professional shops. If it does have significant faults, it will require professional services. While doing the work yourself is interesting and rewarding, and I fully encourage you to try, if you wish, I will warn you that you will spend a lot of time and more money doing it yourself. I will cover the main areas that will need to be addressed to rebuild a starter or generator.

The main areas of concern with any worn generator or starter are the brushes, the commutator, and the bushings. First, the entire unit will have to be thoroughly disassembled, cleaned, and inspected. Brushes will need to be replaced if they are more than half-worn. With both starters and generators, the commutator will often need to be turned down on a lathe and then the segments between the commutator bars need to be undercut. In both a generator and starter, the armature and the coils should be fully tested for electrical integrity. It is beyond the scope of this book to fully explain the procedures for doing this, but primarily there are three tests for the armature: You will want to make sure there are no open circuits or short circuits, and you need to make sure no circuits are shorted to ground. Testing for short circuits between windings is done with a device called a growler. Shorts to ground are found using a test bulb between the commutator segment and the armature shaft. Open circuits are also found through a growler. Field coils should pass a visual inspection and each coil should register approximately 2–3 ohms during a resistance test—5–6 ohms for a 12-volt generator. With starters, be sure to note how the field coils are wired, and test all coils directly at both ends of each individual coil. Any bushings and bearings the generator's main shaft rotates in will usually be worn and will need to be replaced. Starters usually turn in bushings only and do not have bearings. In a generator, the bearings and dust seals often will need to be replaced.

Work on your starter will also include service and restoration of the starter's bendix drive. This is the device at the end of the starter that engages the teeth of the flywheel and retracts when the engine starts. Bendix drives usually require a thorough cleaning, and often a spring with the drives will need to be replaced. You can perform both services and I encourage you to do so. The starter may also need to have its solenoid or switch restored. If the switch is manual, then a simple inspection will tell you if you need to replace it. The posts and main copper bus blocks should be complete and clean. The linkage should be smooth and very little side-to-side motion should be noticeable in the linkage itself. Replacement is the only restoration. The electromagnetic solenoids should make an audible click when current is applied. If not, the solenoid is faulty.

The last item to consider is the generator voltage regulator. The generator voltage regulator can often be cleaned, adjusted, and restored. Again, it is outside the scope of this book to teach this and considering the cost (a brand-new voltage regulator is less than $35 U.S.), it would be prudent to take the voltage regulator in to the same shop you take your generator to and let them test it. If the regulator is in decent shape, they will clean and adjust it if possible; otherwise, they will advise replacement.

WIRING

Restoring the wiring of your tractor will probably be necessary when you do an engine rebuild. Most of the wiring will have dry rotted insulation, patch pieces crimped or spliced in, or will have had sections replaced with inferior

Testing magnetos and distributors often means observing the intensity and condition of the spark created. Creating a test spark plug is an easy way to get that spark where you can observe it, all the while using spark gaps that match your tractor's. Drill a small hole in the base and tap it. Using a small bolt, attach a ground clamp to the plug. Set the gap on the plug, clamp it to ground, connect to your ignition, and you are ready to turn over the engine to test. Be sure to ground the plug well away from a fuel source.

or inappropriate wiring. I once bought a tractor partially wired with lamp cord! Since most of the wiring will be associated with the engine, replacing it during an engine rebuild makes good sense. First, see if your service or owner's manuals have a wiring diagram you can follow. Following what is on the tractor is not a good plan if you are not experienced enough to catch the faults of previous owners. Many manuals will not have a wiring diagram, and making new wiring will require that you go to an antique tractor show, find a restored example of your tractor, and use its wiring design. Be sure to note color, and ask the owner about wire size if he or she is present at the show. Another resource is tractor dealerships or tractor

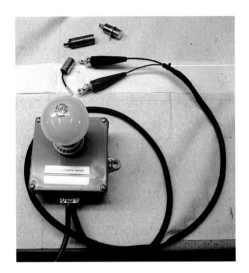

This is a tester made by Weaver's Magneto Repair to test the condensers of their customers' magnetos. While testing condensers using a multimeter is sufficient for quick field checks, fully rebuilding distributor and magneto ignitions requires full testing that loads the condenser similar to the actual conditions.

supply businesses. They usually have special shop diagrams or an experienced hand that may be able to lend or sell you a diagram.

Once you have a wiring design for your tractor, follow the recommendations for wiring color and sizes. If you do not have any ideas on color and size, just remember to use colors that are clear and understandable, and use 10-gauge wires for heavy-duty supply circuits, such as the wire supplying current from the generator to the remainder of the wiring system. Other circuits can be wired with 14-gauge wire. I do not recommend any smaller wiring, but sometimes you will see it used. For accurate restorations, you should use lacquer-coated cotton braided wiring on any tractor made before 1955. After that, vinyl wiring started coming into use,

and whether you should use this or the cotton insulated wiring depends on the make and model. If historical accuracy isn't important, use modern automotive wiring.

Creating the wiring is fairly easy. The only part that is not common sense is the making of the connections. The connections can be made using modern crimping connectors. For historical accuracy, you can use the older twist-on style that must then be soldered into place. This older style requires that a rubber insulating boot be slid onto the wire before the connection is made. Afterward the boot is then pulled over the shank of the connector to help protect and electrically isolate the connector. These rubber boots are difficult to work with, and I prefer using heat shrink tubing instead. Modern crimping connectors require the use of a crimping tool, which is not too expensive to obtain. Occasionally you will see folks use pliers for this task. Emergency field repairs can be made using a pair of pliers, but they shouldn't be used regularly for crimping connectors. Both the old and new style of connectors are made in a variety of sizes and styles for each size of wire and type of connector needed. Spade, or fork, connectors are used whenever the connector is attached to a clamping style terminal. The terminals on a voltage regulator are examples of this. Otherwise, for electrical integrity and assurance, the connector will stay on the post or terminal; if the clamping nut loosens, use the ring, or full-circle style of connector.

Modern connectors are color coded for the proper size. Be sure to use the proper-sized terminal for the wire. Otherwise an unsatisfactory crimp will result. The old style connectors have a size designation on them as well. These are assuming cotton-insulated wiring and will be too big for the same size vinyl wiring. The twist and solder connectors have holes that are threaded. These are twisted on the wire and then the connection is made to the connector by soldering the exposed tag end of the wire to the connector. Heat shrink tubing, or rubber tubes (slipped on before the connectors are attached), are then placed in their final position over the shank of the connector and the wire. Modern crimp connectors already have a plastic boot for the connector and no additional protection is needed.

Spark plug wires should be made at this time. Unlike automotive wires, antique tractor wires are not premade and you must make them yourself. Carbon core wiring is available but should be avoided. Use stranded, copper-cored plug wire. The connectors are crimped on the wire, though not much force is necessary as when crimping modern connectors. Use just enough to dig the teeth into the insulation. Again, like the regular wiring connectors, historically accurate connectors are available. The older style is called a well tip and is recessed so a tag end of the interior wiring can be soldered into place. The modern ones are just crimped.

CHAPTER 15
Drivetrain Repair and Restoration

Drivetrain repairs and restorations are probably the most technically difficult procedures to perform on your antique tractor. Depending on the type and extent of the worn or damaged and broken parts, it can also be the most expensive. For these reasons, most antique tractor restorers perform as few transmission or drivetrain procedures as possible when they restore their tractors. On the other side of the coin, you do have some collectors who restore their tractors' drivetrains completely, being sure to replace all bearings, inspecting every part and reassembling, making sure all tolerances are met. They do this as a matter of course, regardless of whether the transmission actually had any serious trouble. Which approach should you adopt? The answer is either one, or something in between, depending on needs, abilities, and budget.

At the minimum, you should inspect, and carefully listen to and analyze the drivetrain for any noises and possible sources of wear. Replace all seals and change the fluids of the various housings. In short, most of your

This shot shows the five basic components of a crawler's carriage: The sprocket that drives the track, the front roller, the heavy-duty spring tensioning system to keep tension on the track, the idler roller at the top to hold up in the track, and the rollers at the bottom that bear the weight of the crawler.

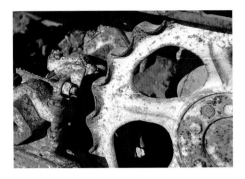

This sprocket and the remainder of the carriage components show considerable wear. Plan on spending a lot of money and time restoring the undercarriage of this crawler.

drivetrain restorations will only address known problems, seal and fluids replacement, and the repair of any other problems or issues your inspections reveal. Most important, be sure you restore and repair the tractor in the most complete way possible that is in keeping with your restoration goals and means.

Inspections

Inspecting the drivetrain will involve jacking up the rear of the tractor completely and very carefully and slowly spinning, by hand, all the wheels and shafts. Do this, looking and listening for any axle, belt pulley, or other part that exhibits a lot of play, noise, or roughness. All wheels, shafts, and pulleys should turn very smoothly without play or noise. Inspect closely for hidden fluid leaks. Many seals or gaskets in a transmission and drivetrain are not exposed, and when these leak the fluid shows up far from the original source. A common leak to follow this pattern is a differential carrier-bearing leak. This carrier seal will leak and then the oil travels away through the axle housing, until it finds a place to drain through. It's important to have a parts manual and a service manual handy while inspecting. This will help you to determine origins and possible causes for leaks. You should also take this opportunity to carefully inspect gaskets. Since you can count on having to replace most or all seals, you should just make a note of the seals you will not have to replace.

Power Transmission Shafts

The various shafts and axles of the drivetrain of your tractor will probably demand the lion's share of your time and resources. These shafts break, have grooves worn into them by seals, and wear at the drive points, such as the splines. Restoring these shafts means you need to remove them, clean them, fill or sleeve worn grooves, have new bearings pressed onto them, and, if necessary, replace them. I will cover the two predominant shaft and axle repairs: rebuilding a seal contact surface and pressing a new bearing.

Over time, the seal around a shaft that keeps out dirt and keeps in oil will wear a groove in a shaft. Not only does this cause a leak, it will also be impossible to repair with a new seal. The new seal will usually hold for a while, but eventually it will allow oil through because of the groove. To determine if the groove is deep enough to warrant attention, just give it the fingernail test: If your fingernail catches on the groove, then it will need to be repaired. Before you fill in the shaft groove, first determine if your new seal will ride on the same precise spot on your shaft as the old one. Often the newer seals are thinner, allowing you to press the seal into the housing to a depth that will change where the lip of the seal rides on the shaft. You may be able to avoid the groove altogether this way. If the new seal will sit on the groove made by the old seal, there is no choice but to repair the groove. The two most common ways to repair a grooved shaft is through building it up with a hard compound, such as a

Antique Tractor Tip

Don't Skimp on Measurement Tools
When shimming components or setting bearing pre-loads, there is no substitute for accuracy. Invest in machinist-quality dial calipers to assist in these chores.

metal-filled epoxy, and placing a bridge over the groove—a process called sleeving. Both methods will work, but sleeving is decidedly easier. Building up the groove with a fill material will be more time consuming but generally cheaper.

To sleeve a shaft, you must first take an accurate measurement of the shaft at some place where it isn't worn. Then take this measurement to any ball bearing and seal supply house and ask for a sleeve. They will probably have one in stock and be able to hand it to you, but they may have to order it. The most common brand of sleeve is called Speedi-Sleeve, which is made by CRC (Chicago Rawhide Corporation). The directions for sleeve installation will come with the sleeve, and the procedure is straightforward. Care and patience will go a long way toward helping get the sleeve on. The sleeve will place a hard steel bridge over the groove, giving your seal a new surface to bear on. The sleeve is so thin that a different-sized seal is not required.

The second method can be accomplished using two different approaches. The first approach uses any easily obtained, inexpensive filling epoxy, such as JB Weld, and filling the groove in with it. You must do this carefully, usually applying two or three light coats. Then you simply sand it down, progressing from a fine 320-grit paper down to something around 600-grit or finer. This will produce a very smooth surface for the seal to bear on. These inexpensive epoxies will wear, and the job will need to be repeated

sometime in the future, depending on how often you use the tractor. If it is a collector's item, this fix may last many years. If you use the tractor regularly, the seal may start leaking in a year or two.

The other option is to use some of the shaft resurfacing epoxy compounds available at the bearing and seal businesses and other industrial suppliers. These epoxies are quite a bit harder and will stand up longer, but typically need to be finished with a machinist's lathe rather than sandpaper. This requirement makes this method fairly expensive and sometimes overly difficult, since the shaft must be removed from the tractor. I prefer this method when there are serious grooves and other nonconcentric damage to the shaft that sleeving cannot repair.

Hi-Low Gearing and Reduction Systems

These transmission systems were designed to dramatically increase torque beyond typical field use. In the early years of these systems, they either used traditional gears with steep ratio gears or designs

This rear drive sprocket shows very little wear. If the rest of the carriage components show similar wear, this crawler would be a good purchase.

based on planetary gearing. They became popular and progressively more common after World War II. This type of gearing is particularly handy for highly variable load conditions in the field where you may need to occasionally increase torque delivery quickly and then move back to the normal range when the difficult area is finished. Getting this torque is as simple as shifting the transmission in and out of high and low. Unfortunately, the planetary style reduction system, such as may be found on International Tractors of the 1950s, can be quite tricky to repair.

Some older antique tractors with Hi-Lo transmissions, such as the Ford N Series with the

Antique Tractor Tip

Bearing Drivers
Use wood or a soft metal such as brass or aluminum to drive or remove bearings. A harder metal, such as treated steel, may damage the bearing.

Track rollers, like the one on this John Deere crawler, require lubrication with general purpose grease every 6 to 10 hours of use; depending on the environment in which it has been used. Crawlers used in wet areas should be greased at the end of the day to force out any accumulated water.

The rectangular bar seen in the upper part of this photo shows how some manufacturers tried to create an adjustable carriage for a crawler. In other words, the owner could push the carriage assembly in or out along these bars to alter the width between the right- and left-side tracks.

Live hydraulic systems were available on tractors until after WWII, and even then it was well into the 1950s before it was common. Many farmers used aftermarket systems to add this feature to their tractors. The Vickers pump featured on this Case is driven off the crankshaft and provides full-time live hydraulics to this tractor.

Sherman retrofit or the Allis-Chalmers W Series tractors, use a secondary transmission with much of the same design principles as the tractor's regular transmission. The newer torque amplification units such as the Farmall unit and the Minneapolis-Moline's Ampli-Torque are examples of a shift-on-the-fly system. Because of the great stress placed on these units and their shift-on-the-fly characteristics, they often will be in a state of disrepair when you buy the tractors. Fortunately, the torque amplification unit is usually only broken in one of its position settings. Usually the tractor will run fine in the other setting. Some antique tractor restorers, knowing that they will not be relying on the tractors for any field work, just leave the unit as is, never using the setting that doesn't work. It is beyond the scope of this book to cover the repair of these units, other than to mention the work is similar to other transmission procedures. They are different from regular transmissions in that the repair will tend to be more involved and more expensive, and parts may be hard to come by.

Transmissions

Restoring the transmission involves disassembling it, thoroughly cleaning the housing and the individual parts, and reassembling, being sure to renew worn and broken parts. In the process, you will also adjust bearing preloads, mesh patterns,

and nut and bolt torque in an effort to bring the transmission back to near-new condition. Probably the biggest areas of concern with most transmissions are worn gears, gears with broken teeth, bearings and races, and shifter assembly. The gears themselves occasionally have broken teeth, but usually the gear teeth are just worn. You must use your own judgment when replacing gears. Gears with broken teeth must be replaced. Highly worn gears ideally should be replaced, but the cost can be prohibitive. The decision is up to you, but I would suggest replacing any highly worn gear in a working antique tractor and leaving alone the gear in any antique tractor simply being collected or used for light work. Often the snap rings and locking plates that keep reverse idler gears positioned are worn. Be sure to replace all of them if needed.

There is a second shaft in the transmission, the drive shaft, which takes a lot of abuse and bears a lot of the torque of the tractor. This is also called the output or bevel pinion shaft. The renewal of this shaft should involve new bearings and races, and the shaft must be reshimmed to mesh with the differential. The bearing preload of the shaft must be carefully set. Shimming the bevel pinion shaft to properly mesh the bevel gear to the differential ring gear and establish bearing preload is an operation that requires some training, and I don't recommend that the novice mechanic perform it alone. I recommend having a more experienced friend help you the first time. I learned the technique working on other equipment and had the fundamentals down pat, and I still struggle to get it right. Making matters worse is the fact that we must sometimes reuse worn components because replacements are cost prohibitive or nonexistent, which may make it impossible to get the adjustments perfectly right. If this is the case, just getting your adjustments as close as possible is all you can hope for, unless you can afford to replace the bevel pinion gear and differential as a set.

While the last paragraph was meant to give inexperienced mechanics pause, these procedures are not impossible. This section should serve more as an overview or background to be coupled with further training, or as a refresher for those who have not done this for a while. If you are with me so far, the first thing we will do is set the in-line mesh pattern. That is, we will adjust where the teeth of the bevel pinion gear touch the teeth of the differential ring gear, relative to the outside and inside of the ring-gear teeth. The differential section discusses where the bevel pinion teeth touch relative to the ridges and valleys of the ring-gear teeth. This second adjustment is called the lateral adjustment.

First, install the bearing on the pinion side of the shaft, being sure to install your shims between the bearing and the shoulder of the shaft or pinion gear. To start off,

Antique Tractor Tip

Shift Mechanism Tactics
Inspect gear shift levers, gear shift lever pins, forks, and rails for excessive wear. These items typically are highly worn, but are easily built back up with careful MIG welding, followed by filing and shaping.

use the shims that were on the shaft when you disassembled the shaft, plus a .005-inch shim. Install the shaft in the transmission case and tighten the end nut of the bevel pinion shaft so the shaft is snug against the bearings. Then install the differential, being sure to completely seat the bearings and the races against the adjusting nuts.

Installing the shaft and the differential doesn't require any precision at this point, just make sure neither is loose or sloppy in its bearings. Then paint the differential teeth with a marking paint, Prussian blue being the compound of choice here. Spin the pinion shaft, looking at

the marks the teeth make in the marking compound. Your tractor's service manual should show you how the marks should look in the paint. If not, you want to see marks that are very close to the center of the teeth of the differential ring gear, but are slightly nearer the ring gear's outside edge. Where these marks are in terms of nearness to the ridge or nearness to the valley doesn't matter at this point; we set that when we set the differential for the last time. If the differential and pinion are highly worn, the marks will be even farther off center. The reason you don't center the

Antique Tractor Tip

Groovy Shafts
When installing new seals on shafts, be sure to check for any grooves worn into the shaft. If the groove is deep enough to catch your fingernail, the groove will most likely cause the new seal to leak. Grooves can typically be sleeved or filled with epoxy.

Here hubs are bolted to the rear wheels of a Farmall M. These hubs are how a second set of rear wheels is attached to form dually rear wheels.

marks on the differential teeth is because once the tractor is placed under a load, the teeth will splay somewhat, leaving you with a perfect mesh pattern. Repeat these steps as necessary, adding and removing shims to establish a mesh pattern on the teeth consistent with your manufacturer's recommendation.

Once the mesh pattern is established, then you must create the proper bearing preload. Some shafts specify a certain torque for the nut at the end of the shaft, while others will actually have shims that are placed in front of the front bearing. In cases where shims are used, the procedure is again somewhat time consuming. The trick is to add enough shims to allow a little end play in the shaft, then measure the end play. You will need to remove as many shims as necessary to eliminate the end play. Then the preload shim is removed. Let's create an example to illustrate this procedure. First, you consult the owner's manual and find the bearing preload is .0025 to .005 inch. Next add a shim that measures between .0025 and .005 inch. This is called the preload shim and will be the shim removed as the last step. Then you add enough additional shims to the shaft to create end play in the shaft when it is installed. Let's say you have added enough shims to give the shaft .005-inch end play. Now that you have end play, you can start removing shims. Since you have .005 inch worth of shims too many, you start by removing .005 inch worth of shims. Next, remove your special preload shim. Voila! You now have a transmission output shaft that has been properly shimmed to establish the proper bearing preload.

Differentials

Rebuilding the differential means renewing the bearings and, if warranted, the ring gear. The ring gear should only be replaced if you can also replace the bevel pinion gear as a matched set. The differential sits in the differential housing with adjusting nuts that provide for the lateral adjustment of the gear lash, or the play, between it and the bevel pinion shaft gear. The best way to set this is to use the same Prussian blue compound we used to set the inline mesh pattern earlier.

Antique Tractor Tip

Match Those Gears
When assembling a drivetrain, consider installing matched gears on the side opposite from where they were removed. For example, bull gears are usually identical, and if you swap sides, you "trade" the excessive gear lash in the forward direction for gear lash in the reverse, and less important, direction.

Setting the lateral mesh pattern simply involves turning the adjusting nuts until the marks made by the pinion gear in the Prussian blue are close to the center of the ring-gear teeth, yet are off-center slightly toward the top of the ridge of the ring gear. This allows considerable gear lash, but once the tractor is under load, as I mentioned before, this lash tightens up.

Final Drives and Bull Gearing

The final reduction gears on tractors are often worn but serviceable. Gear lash is often not adjustable; however, I find these gears are usually reversible between the right and left sides, and if the pinion gears that drive these bull gears are reversed as well, the wear profile in the forward direction is very minimal. Yet reverse will now have the badly worn profile that forward did before you switched them. Since reverse is almost never used under load, this is an acceptable trade-off. As was the case with the transmission, replace any gear with broken teeth and look to replace most of the seals and snap rings during the renewal.

Brakes

Brakes are simple to renew and restore. Both disc- and drum-style brakes can be relined by any full-service automotive brake renewal company. It is often rather expensive, and many folks try to save the money by installing their own linings. If you are comfortable with your abilities, you can give it a try. I never have relined the brakes myself, instead opting for a local shop. This shop uses the old style lining, but uses modern bonded lining technology, which is much safer and more reliable. Bonded lining is something most folks cannot do in a home shop. If you want to reline your brakes yourself, the procedure basically involves setting brass rivets to fasten the lining to the brake bands and discs. Heading the brass rivets must be done 100 percent correctly to keep the brake lining from peeling off the band.

The next step in brake rebuilding is turning and truing the drums and friction plates. This can be done by any competent machine shop. Often one of the friction-bearing surfaces of the disc-style brakes is the differential or transmission housing. In this case, simply using a large flat milling file to flatten and true the surface will suffice.

Many parts of an antique tractor, such as transmission, final drives, and combined hydraulic/transmission compartments, require oil. A lot of oil might be needed to fill the compartment after draining. These lubricants are therefore sold in 5-gallon containers and some sort of bucket pump will be needed. These pumps are handy for merely topping off these fluids. Older bucket pumps, such as this one, last much longer than the newer cheap plastic ones. Trying to find long-lasting durable tools such as this one is another reason why you should frequently hit the farm sales and auctions.

Antique Tractor Tip

Belt Work Tractors
When looking to buy a parts tractor to fill the need for replacement transmission components, guessing the condition is difficult and inspection of internal transmission parts is nearly impossible. In this situation, try to find a tractor whose history indicates a lot of stationary belt/PTO work, because the gears, bearings, and shafts of the drivetrain will be fairly fresh.

The rear wheels of tractors are made up of three parts: the rim the tire sits in, the disk that provides the support for the rim, and the hub, which is an integral part of the rear axle (or bull gear shaft). The disc is often rusted to the hub and can be tough to get off. The easiest way to remove the wheel disc from the hub is to jack the rear wheel off the ground, leave the tire and rim in place, and remove all but two disc-to-hub bolts. Then slightly loosen the two bolts. Using the leverage the tire and rim affords you, begin rocking the whole assembly until the disc frees itself from the hub. When this happens the two bolts will catch the assembly and keep it from falling to the floor, or even worse, on you. For this reason, never take all the bolts off at once until the assembly is loose and you can enlist the help of a hoist or a friend to help remove the wheel assembly. This is especially important if the tire is loaded with water.

Your final checks during brake reconditioning are making sure that set screws are adequately tightened and secured, that all brake rods, cams, and brackets are tight and clean and without welds or other defects, and the brakes are adjusted. Your service manual will outline adjusting the brakes for your tractor.

PTO and Belt Pulleys

PTO shafts and belt pulleys are like the other systems of power transmission of your tractors. Restoration requires working with lots of gears, bearings, snap rings, and adjustments. Among the most common repairs needed for the PTO during a restoration are replacing the outboard bearing of the shaft, replacing a broken PTO or belt-pulley shaft, and sleeving these shafts. The

shaft of a continuously running PTO is usually in dire need of sleeving, as the constant running under a seal usually wears a significant groove.

Whether your tractor has a "live" style PTO or a typical transmission-driven PTO determines what issues and subsystems you need to address. If your PTO is transmission driven— that is, driven by the input shaft of the transmission—then your rebuilding concerns are limited to the PTO assembly itself, possibly a pilot bushing in the input shaft, and the linkage to engage the PTO. If the PTO is a live or continuous-running PTO, then you must inspect and restore the secondary drive system that drives the PTO. Two designs of live PTO are common, and one of these designs has two variations. The truly continuously running PTO has its own driveshaft that bypasses

the engine clutch. This typically involves some type of driveshaft within a driveshaft, like the Co-op tractor that brought the idea of a continuously running PTO to fruition. The second main type of live PTO is sometimes called the Run on Demand PTO, which is live and separate from the transmission, but the PTO can still be stopped through a clutch. This is either a two-stage engine flywheel clutch or a separate clutch, usually a hand clutch, at the transmission.

Belt pulleys also require attention and inspection during a rebuild, and much of the work and issues are identical to those you will encounter during PTO restoration. The pulleys themselves, if they are made of paper, can be restored by several reputable firms across the country (see the supplier and vendor section). The assembly as a whole is typically either very worn— because the tractor stayed belted up to a load for years and years— or is hardly worn at all because belt pulleys otherwise are used infrequently. While roller bearings are commonly used, finding cup and cone style bearings is common in belt-pulley assemblies and setting bearing preloads is a chore you may run across. This will be straightforward and should be outlined in your service manual. Usually it is set using gaskets as shims under a plate that holds the front bearing in place. I have also found that belt-pulley housings will often have very large freeze plugs that will require removal and setting.

Painting Your Antique Tractor

One of the most rewarding experiences during an antique tractor restoration is standing back and admiring the finished tractor. It sits there, looking for all the world like it just came off the showroom floor and suddenly you realize you are looking at a piece of history. You have rolled back the clock and have been given a window on the past.

Another wonderful feeling that comes from a well-done paint job is the realization that you have completed the restoration. After a long restoration, it is a good feeling to be done. Sure, there are decals to apply, buffing, and polishing in addition to correcting some of the minor problems that come up with any restoration, but finishing the painting means you are in essence finished.

If personal satisfaction isn't enough to convince you to paint your own tractor, the cost of a professional paint job probably will. If you take your time, read this chapter, and enlist the help of an experienced friend or community college auto painting course, you will have a paint job that will be everything you dreamed it could be.

Rust is a very common problem found on most antique tractors, but this is worse than usual. Rust compromised the strength of this fender to the point that it cracked. The remainder of the fender is highly pitted and will probably need to be replaced.

One of the handiest and easiest ways to de-rust smaller parts is to use a process called electrolysis. Basically, you use a small battery charger to create a chemical reaction that changes rust to other compounds. There are many different sites on the Internet that outline how to do this safely, easily, and effectively. However, there are some safety precautions, so be sure to do all the research beforehand. I find most of the sites don't mention that you can't use the float or automatic setting on the charger. Your charger must have a constant charge setting or you'll need to find another that does.

Cleaning an Antique Tractor for Painting

The first step to painting a tractor is making sure the paint you apply will stick to the tractor. Simpler words were never spoken that in practice were harder to accomplish. Since how well you strip the tractor of rust, paint, and grease determines how well your paint job will look and how long it will last, you should pay particular attention to preparation. Thomas Edison is credited with proclaiming that the process of invention is 1 percent inspiration and 99 percent perspiration: Well, I am here to say that painting is 99 percent surface preparation, 1 percent paint application. While that might be a stretch, I can unequivocally state that nothing has more to do with the quality of your paint job than surface preparation.

Unfortunately, all sorts of obstacles will stand in your way: Any trace of grease, rust, weak old paint, no matter how slight, will cause problems. Any wrinkle or dent will stand out like a sore thumb. To make sure you apply a finish to your tractor that will make you proud and last for a long time, you have no choice but to try your best to remove every last bit of rust, paint, and grease from the tractor. Considering the copious amounts of all three of these on antique tractors, it should be painfully obvious how difficult and time consuming surface preparation can be. The task is more bearable if you divide it into manageable mini-projects that are interwoven with more enjoyable tasks. Regardless of your approach, if you avoid taking shortcuts and remain patient and thorough, you will end up with a surface that will hold paint and look good for an extended period of time.

First, you will need to decide how far you will disassemble the tractor before you start surface prep. This is a personal decision based on your own needs and desires, but the further you disassemble, the easier the parts and tractor are to clean and paint. You will more than likely clean and paint each part more thoroughly if you remove it from the tractor. The main reason this is true is because every area of contact between parts can harbor rust and dirt that will prevent paint from sticking to the metal. Over a period of several weeks or months this can

Antique Tractor Tip

Grind, Brush, Scrape, and Strip
When cleaning a part for painting, use a combination of approaches to get the part in the best possible condition. For example, clean off the worst of the grime and dirt with a wire brush, then strip paint and the remaining grease with a lye bath, and remove the last stubborn bits with sandblasting.

cause any new paint you apply to lift and peel. I recommend disassembling the tractor as completely as possible before beginning cleaning and other prep chores.

The next decision will be how you will clean and prepare the tractor for bodywork and paint. I wish I had a standard answer for you here, but you will probably need to use a variety of methods to clean and strip the tractor. The methods come down to sandblasting, various hand-held power tools, manual tools such as scrapers, chemical paint strippers, rust removers, and degreasers. Each has its own advantages and disadvantages, and there are times when any one method should not be used for either safety reasons or a risk of damaging parts. I will cover all the ways you can prep your tractor along with some notes on safety, effectiveness, and the appropriate and inappropriate times for using the method.

SANDBLASTING

Sandblasting is usually the quickest and easiest way to remove old paint, dirt, rust, and scale. It has several drawbacks, however, and is not something you should do to every part of the tractor. Many people swear that sandblasting an entire tractor in an assembled state is safe if you are careful to seal holes, seals, joints, and so on with duct tape before sandblasting. I personally refuse to sandblast any part of a tractor that contains precision assemblies. I only sandblast individual parts that I can remove from the tractor. I am not sure why folks seem to think sandblasting an

assembled mechanical device is safe, as I have seen sand work its ways into housings and assemblies through holes and gaps never noticed, and I have seen one transmission that required disassembly because the owner sandblasted the transmission housing. How the sand got in was a total mystery to him because he said he taped closed every hole in the housing.

My recommendation is don't sandblast an assembled precision component. The risk just isn't worth the saved work in my judgment. Sandblasting is covered in a little more detail later.

LYE-BASED CLEANERS

Lye-based cleaners are very effective, very cheap (if you make your own), and have many benefits. Most parts of a tractor, with enough ingenuity, can be soaked in this cleaner. The real benefit is that lye does two jobs at once: The solution completely degreases the part (though I remove excess grease manually first) and if the paint on the part is not a modern caustic-resistant paint, the lye solution will completely remove any trace of paint. The added benefit to this method is that the really nasty

Use wire wheels installed on bench grinders for removing rust and scale and dirt from parts. They will also bring out the shine in metal before you paint or install them.

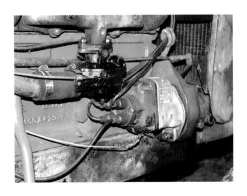

Restoring an antique tractor is a large project. Sometimes it makes more sense for busy folks to complete the job in stages. This engine, carburetor, and magneto were rebuilt; and then the engine was painted and installed back on the tractor. The rest of the painting job will have to wait until the owner has more time available to finish the project. This approach is preferable to the other approach, which is the tractor, in pieces, sitting idle for years.

Antique Tractor Tip

Saving Sand
You can reuse sandblasting sand a few times before having to discard it. Try to sandblast on concrete pads or similar surfaces where the sand can be swept up for reuse.

When painting an antique tractor, stripping the tractor down to its bare frame is usually easier to clean and paint and generates betters results in the long run. Trying to clean and paint around installed engines, steering components, and the like generates almost the same amount of work as removing the systems and parts and separately cleaning and painting them. This Case V model tractor has had its frame painted and is waiting for its engine to be painted.

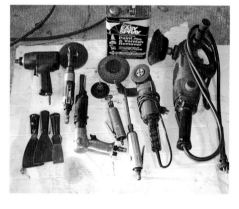

When you decide to paint your old tractor, the first hurdle is: How do I get rid of the old paint, dirt, and rust? I have shown and discussed many sandblasting options, but the most common approach involves these tools and some elbow grease. Here is a wide assortment of air and electric wire wheels, grinders, scrapers, needle scalers, and a chemical stripper. You should always wear a dust mask when mechanically stripping paint. Most paint on antique tractors contains lead.

Preparing an engine for sandblasting involves careful work to make sure there are no cavities or holes that would allow sand to work its way into the engine. Duct tape holds up well to sandblasting pressures, so packing holes with crumbled paper and covering them with duct tape is a good strategy. Holes that don't have an adequate surface for duct tape can be plugged with expandable pipe plugs, as seen here in the distributor hole. You are then ready to suspend the engine and begin sandblasting.

toxins from the grease and paint, like heavy metals (old paint is lead based!!!), all end up in the bottom of the tub. If you were to sandblast or use a wire brush to remove the grime and paint, these toxins would become airborne. For this reason alone I recommend soaking all the parts possible in a lye-based cleaner.

MECHANICAL REMOVAL

Some of the mechanical methods are handy for removing paint, grime, and rust from a part. This is especially true if you prefer not to sandblast the part and it is too big to soak in the lye solution. I usually use a wire brush attached to an electric angle grinder to remove paint and rust from these

types of parts. I will tell you now not to bother trying to use a wire brush attached to an electric drill as a substitute. It doesn't spin fast enough to remove paint and rust with any expediency whatsoever. If you prefer mechanical stripping to chemicals, I recommend investing in a nice angle grinder. I promise you won't regret it. You may also be able to rent one.

Another tool for this purpose is called a needle scaler. It is pneumatic and operates by beating an object with a set of hardened steel pins. This tool appears to be effective at removing stubborn paint and scale (rust transformed into a slaglike compound, often called rust weld), and moderately effective in removing rust. It makes short work of cleaning hard-to-reach and curved surfaces. Rest assured that no matter the combination of methods, tools, and products you use to clean and prepare your tractor for bodywork and paint, the more thoroughly you clean, the happier you will be with the end results. I haven't tried a needle scaler, but a friend swears by it.

PRESSURE AND STEAM CLEANING

Another tool to rent that you might find handy is a steam-powered pressure washer. These machines spray hot steam under extremely high pressures, doing a great job at removing grease and loose paint. They can also be used in conjunction with degreasers to enhance the steam cleaner's effectiveness. Pressure washers, with or without chemical additives, also do a great job of cleaning tractors. Be careful of electrical components when pressure or steam washing a tractor. These two methods are particularly effective for use at the front end of the restoration, removing all the loose dirt, leaves, and extra grime from the tractor.

MANUAL METHODS

Old-fashioned elbow grease, sandpaper, oven cleaner, chisel, and putty knife also make an unbeatable team for effective, thorough cleaning. If a tight budget won't allow investment in tools, then this is the method of choice. Unfortunately, they involve lots of time and hard work and while slower than all these other methods they are perfectly appropriate. If you don't try to rush, these methods have a therapeutic quality that many people find enjoyable. Hand-scraping the tractor also isn't as difficult as it seems if you use a solvent or oven cleaner on the surfaces beforehand to soften the caked-on grease and old paint. Another big advantage to this method is it can be noticeably safer, as the toxins in old paint and grease don't become airborne. The trick to making this method as effective as possible is making a small assortment of your own custom scrapers from scrap steel stock and keeping them sharp as you work.

Before you paint an engine, it will first need to be thoroughly cleaned. Suspending the engine on a hoist makes all of this much easier. Sandblasting is the easiest way to clean the engine, but you'll need to make sure every possible hole is closed off so the sandblasting abrasive can't enter the engine. Some of the things you'll need to do to achieve this is to pack coolant hose inlets and outlets with paper and masking tape, fit bolts or old sensors in the coolant and oil pressure sensor holes, check the condition of gaskets and replace as necessary, and pack thin ropes of paper in and around the rear and front main seals if necessary. Brass fittings are aggressively eaten by the sandblasting material, so be extra careful around these parts. After sandblasting, you are ready to begin painting. Be sure to use a primer rated for bare metal and begin applying it in consistent thin layers. On pieces full of nooks and crannies, like an engine, it's often easier to paint these difficult areas first, then work out to the broad flat areas. Continue by applying the paint. Build the coat in several passes until you have a solid coat of paint that neither sags nor runs yet looks full and wet.

THINGS NOT SO GREAT FOR CLEANING AND PREPPING A TRACTOR

Although they can perform the job of removing paint, I avoid using chemical strippers. The primary reason for this is safety. Traditional paint strippers are very toxic and difficult to handle. The fumes are bad for you and they can damage your skin or eyes. Never use them without adequate ventilation.

Chemical strippers are also time consuming to use since it can take several hours and more than one coat to completely strip all the paint. In addition, you must be sure to remove every last trace of stripper afterward or the new paint you apply will be stripped off also! Of particular nuisance is the tendency of chemical strippers to hide in the cracks and crevices of a tractor, causing new paint to peel off at these seams later.

Finally, chemical strippers are not cheap. If you buy enough to

Whenever you begin a painting session, you'll need to strain your paint. Straining prevents any small, hard dry particles from the lid or other contaminants in the paint from clogging your spray gun or ruining your paint job. Paper strainers are disposable items and are often given away by your paint supplier. Be sure to get a bunch when you buy your paint. Some primers are not thinned or used with additives and can be returned to the paint can. You should strain the paint again when you return the remaining paint into the can.

any circumstance you run across when restoring and repairing an antique tractor. There is no reason to use dangerous fuels as cleaners anymore. I suggest taking a look into these newer products.

Sandblasting

Sandblasting does a great job of getting rid of rust and old paint, and even light caked grease. There are some things you should never sandblast. Never sandblast any polished surface or any soft metals, plastic, Bakelite, and so on. Sandblasting can ruin these materials if you aren't careful. Parts from certain systems like hydraulic pumps and cylinders, internal engine and transmission parts, and radiator cores and tanks should also not be sandblasted. The risk of damage is just too great. Some housings, typically the rear-end housings, that ordinarily would be good candidates for sandblasting after complete disassembly, often have a tarlike sludge on the interior surfaces that must be removed before sandblasting, or the sand will just stick to it, forcing you to clean the part completely after sandblasting. As I mentioned earlier, I also do not sandblast any assembled systems or portions of the tractor because sand always seems to work its way into places where it will cause problems. Beyond these warnings, use your common sense. Most things on an antique tractor can be sandblasted, at least if the part is removed from the tractor first, and sandblasting is a very effective, powerful tool for the restorer.

do a whole tractor you are talking about some serious money. The strippers that are nontoxic are particularly expensive.

The other cleaner I didn't mention, and which should not be used, is gasoline. Never use gasoline as a cleaner! It is highly flammable, the fumes are toxic, and it isunhealthy to have on your skin. What's more, gasoline isn't even as

effective as the lye-based cleaners. Heavier fuels such as diesel and kerosene are a little more appropriate as cleaners, but still require adequate ventilation, are dangerous, and are not as effective as modern degreaser compounds or lye-based cleaners. There are many different degreasing compounds available at your auto parts store that can be used on any part or in

When restoring a tractor, you should count on buying a small siphon style of sandblaster. These

Antique Tractor Tip

Sandblasting Caution
If you don't remove the transmission and engine before painting, use wire brushes, a pressure washer, and lots of elbow grease to clean the exterior. Sandblasting the assembled engine or transmission can ruin seals and any sand forced into the cases will damage moving parts.

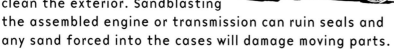

are not very fast or effective, but they are inexpensive and do the trick for most of us. They are great for small parts and work well with a small sandblasting cabinet. Faster, more effective types are known as pressure style sandblasters (also called pressure pots). These, coupled to a high capacity air compressor, really make short work of sandblasting. They are not necessary, except when you are ready to sandblast really rusty, large pieces like wheels and so on. The pressure pots are available at tool rental yards. Since they are only really necessary a few times during a tractor restoration, I would recommend hiring out this work or renting a high-capacity pressurized sandblasting system once you've isolated the most difficult components to strip.

There are many different types of abrasives you can use with sandblasters. While sand is the most common material, there are other media, such as aluminum oxide, glass beads (handy for some types of delicate parts), walnut shells (handy for engines and transmissions where the shells pose no threat to the internal workings), and others. I use sand exclusively simply because the other media are too expensive. I use two different grits, fine and extra fine. Coarse for some reason is completely unavailable locally except as washed river sand. This sand is very dusty, and I do not use it for that reason. I can mail-order it, but I find it unnecessary. Medium is never available in my locale, either. Fine seems to cut through the worst of the rust and paint effectively

---

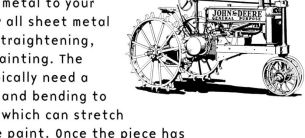

Antique Tractor Tip

Sheet Metal Tricks
Mount sheet metal to your tractor after all sheet metal repairs and straightening, but before painting. The piece will typically need a little pulling and bending to fit properly, which can stretch and bend the paint. Once the piece has been fit to the mounts, you can take it off and paint knowing it will not need to be bent further.

---

and extra-fine does well with sheet metal, taking off rust and paint without overheating the metal or removing too much metal.

When sandblasting, move the nozzle around the surface of the piece constantly until all remnants of rust, grime, and old paint are removed. Sandblasting is best done with a side-to-side motion that does not tarry on any one spot too long. Revisit spots later that sandblasting does not clean up in the initial pass. This is particularly true when sandblasting sheet metal with a high-volume, high-pressure sandblaster. Sandblasting with these heavy-duty sandblasters can generate enough heat in the metal to warp it. Never work on one spot on the sheet metal too long. Once you have the piece mostly clean, inspect it for any other problems that may have been hidden by paint and rust.

Sheet Metal Repair

When someone thinks of a restored antique tractor, they think of a sleek, new-looking tractor, complete with smooth lines, great-looking paint, and wrinkle- and dent-free sheet metal. Unfortunately, returning sheet metal to like-new condition is time consuming if you handle it yourself, and expensive if you hire the work out to a professional. This is the reason most collectors insist on buying tractors whose sheet metal is in excellent shape. Even with a relatively clean tractor, you will probably have to straighten some of your sheet metal. You start by beating out dings and dents using a standard bodyman's swirling hammer (shrinking hammer) and dolly. Mallets, ball-peen hammers, and even small hand sledges can be used too. This procedure is called bumping and is easy to learn, but it can take years to master.

How, exactly, should you go about doing this? The short explanation is to beat the metal back into shape. The long explanation would require volumes, but I'll point out a few things. Simply beat on the sheet metal to repair

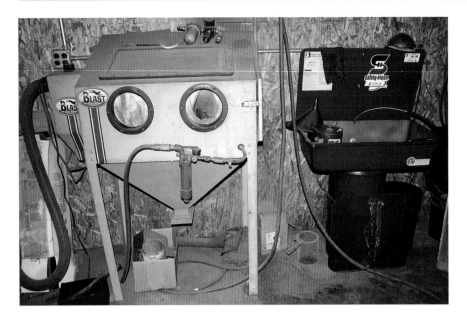

Thorough restorations mean painting, and great paint jobs start with parts that are as clean as a whistle. Here are two tools to make the work of cleaning parts quick and easy: A sandblaster and a parts washer. The unit on the left is a Skat Blast sandblaster and is considered one of the better brands on the market. The parts washer on the right is a Safety-Kleen barrel-style parts washer and is also considered an excellent brand. Barrel-style washers are usually preferable, as they store a lot of cleaning solution and leave a large open basin for washing large parts. Be sure to use the safer water-based cleaners in your parts washer. While mineral spirits-based cleaners are slightly more effective at cleaning, these cleaners are a fire hazard that part-time hobbyists do not need to risk using.

the dent or ding, using care not to stretch the metal, and use heat whenever possible to help with the straightening process. You can use a welder's oxygen-acetylene torch. The idea behind this bumping is to get the piece straight and restore the lines and style of the metal. Your next step is to make sure the sheet metal lines up with the mounting brackets on the tractor. You may have to do some more pulling and straightening to achieve this. If it doesn't line up,

use controlled ways (such as pushing by hand, pulling with a come-a-long, and so on) to slowly bend and push the metal into the required shape. Make sure the frames and brackets for the sheet metal are straight first!

If after all this bumping, beating, and straightening you have a section of sheet metal that is not smooth, it is likely that the metal was stretched either by the original accident that caused the dent, or the section was stretched during the process of repair. One trick for repairing this type of problem is to "shrink" the metal. This is done with a heating tip on an oxygen-acetylene torch. Heat the area until it glows with a dull cherry red, and then rapidly quench it with water. This has a tendency to shrink the metal and may help remove the worst of the waves and wrinkles. If not, you may have to cut out the section and weld in a new piece.

After all of this, you are ready to actually replace the sections that cannot be bumped or straightened. We repair dings and dents first because cutting patches to fit is much easier if the form and shape of the sheet metal is right before patching. This way your patch will be the perfect size and orientation and will have all the right lines and curves. Also, bending and orienting a piece of sheet metal is more difficult after welding in patches. Be aware that you may have to do some manipulation after patching because the heat from welding the patches can cause the metal to warp.

Installing patches is easy once you have learned some basic MIG welding skills. Simply choose an

Antique Tractor Tip

Bondo Lightly

Use of plastic filler to hide blemishes in metal should be minimized and used only sparingly. If you find yourself applying it in coats thicker than a few millimeters, you should consider repairing the area more thoroughly first.

outline for the patch in the original sheet metal, using permanent marker. Cut along these lines. This will leave clean, solid edges along which to weld the patch. Cut out a paper or cardboard template to fit the hole in the sheet metal you have just made. Use this template to cut a patch out of a scrap piece of sheet metal that is the same thickness, or gauge, as the old sheet metal. Typically antique tractor sheet metal is about 16 gauge. Use tape to hold the piece in place until you have made a couple of small tack welds. Remove the tape, welding the piece at several points, being sure not to let too much heat build up in one place. Many restorers simply spot weld the piece in, skipping about an inch between spot welds. Grind away excess metal at the welds. Later on, during the finishing phase, body filler will fill in the gaps between the welds as well as grinding marks and any pinholes caused by welding.

Now that you have a piece that is sandblasted, cleaned, patched, and straightened, it is time to make any brackets and other pieces that are necessary for installation. Work the sheet metal and the brackets, making final adjustment to fit and finish so that everything is pleasing to the eye and hand. One experienced bodyman I know closes his eyes and uses his hands. He says he can feel imperfections in the metal that his eyes do not always see. Very minor imperfections that could be corrected with 3/16 inch of body filler or less may be ignored at this time. We won't apply it now unless for some reason you do not plan on priming the metal. You can apply it after etching the metal with phosphoric acid if you are not going to apply primer to the metal.

Applying the Finish

INTRODUCTION

Now that the tractor is clean and free of previous paint and grease, you still have some more work to do before you start spraying your finish paint. Sheet metal has to be straightened and repaired; the tractor has to be thoroughly wiped with a degreaser one last time; and brackets, fasteners, and other odd parts that you never have been able to get around to obtaining have to be obtained. Parts have to be test fitted, and various parts of the tractor have to be suspended in preparation for painting. You have to shop for paints and finishes, and now would be the time to take a quick auto painting course at the local technical or vocational school. In other words, cleaning the tractor to reach this point in the restoration was a large part of the job, but there is still plenty of work to do. Doing all this in preparation for applying the paint is time consuming, and at times tedious and unexciting. Rest

While it would be great to spray paint directly from the can without additional work, you'll almost always have to mix the paint with a thinner (more properly known as a reducer) and possibly some additional additives you've chosen. Here is a gallon of paint shown with a matching reducer with an average or medium rate of evaporation and a hardener. The hardener will increase the ding and chip resistance of the paint and add additional gloss to the paint.

When restoring your antique tractor, the paint you use doesn't need to be anything fancier or more expensive than acrylic enamel. The close-up of the label on this gallon of paint (tinted and mixed to match Minneapolis-Moline Prairie Gold) will give you a better idea as to what to look for and expect from your paint dealer.

Antique Tractor Tip

College Pays
Enrolling in local community college painting and autobody courses will hone and sharpen your skills. A welding course should be considered as well.

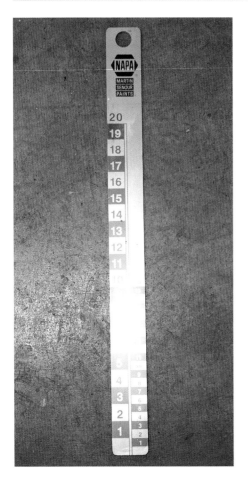

Paint mixing recipes are usually expressed in parts, as in "x parts of paint to y parts of thinner." Here a gradient stick is used for mixing paint. For example, 5 parts paint to 1 part thinner is easy to mix by placing the gradient stick in the mixing cup and pouring in paint to the 5, 10, or 15 mark, and then adding thinner to the 6, 12, or 18 mark. The side with smaller gradients is for smaller batches and the other side for larger batches.

assured that just like the cleaning phase, diligent, thorough work during this phase of your tractor's face-lift will reward you with a lasting finish that will protect the tractor, enhance its value, and be a source of pride. While applying paint is within anyone's ability to learn, mastering the art is difficult, and taking the time to prepare and paint a few pieces of scrap sheet metal will help you learn some of the nuances of painting.

Safety is a critical issue with painting. Never paint or mix paint without a respirator. Make sure your respirator is a proper painting respirator. If you aren't sure, ask your paint supplier. If you decide to use modern paints containing isocyanates, paint in a fully ventilated area or in a fresh air painting suit. Isocynates can cause a reaction similar to an allergic reaction that will cause you to pass out, become sick, or even die. Once you react to these substances, you can become "sensitized," meaning that you will react violently every time you are exposed. Thinner, solvents, and other additives are equally dangerous. Most are highly flammable; do not even think of

painting around an open spark. This includes painting in a shop with an exposed electric motor such as a compressor motor, an open flame stove, tobacco use, or any form of open ignition. All modern paints come with a material safety data sheet. If your paint supplier doesn't offer you one, ask for it. This sheet will outline all the precautions you will need to ensure safe use.

APPLYING THE PRIMER AND PAINT

Applying the primer and paint requires the use of a spray gun and a source of compressed air. Fortunately, spray guns have very minimal air requirements, typically within the capabilities of any small, used air compressor. Spray guns vary widely in quality and price. Like most things in life, you get what you pay for, and you should consider purchasing a quality gun if you can afford it. Companies such as Binks, DeVilbiss, and others make nice guns. Regardless of the type and style, all guns must have the air supply filtered for water and oil. The heating and cooling of the compressor and compressor tank causes water to condense in the tank, which can then enter into the air supply. "Drying" the air involves installing a water trap in the airline. I use two, plus I have a contaminants filter at the gun. If your compressor is overfilled with oil, oil can make it into the air supply. Be sure to have the proper oil level in your compressor.

What style of gun you need depends on your own preferences, and your local store should be able

Antique Tractor Tip

Paint Systems

When deciding on paint and all the additives to use, consider using an entire system from a paint vendor. These painting systems consist of paint and additives guaranteed to work together.

to assist you with the various styles. The most flexible and trouble-free style is the siphon-fed, nonbleeding cup storage air gun. This means the paint is fed into the gun, creating a siphon effect that draws paint into the air stream. A bleeding style gun means air is constantly moving through the gun, even when it is not in use. Other styles are available, but be sure to find one that you are most comfortable with. There are many sizes available, and if I had to settle on one gun for tractors, I would choose a touch-up size spray gun. Quality touch-up spray guns are smaller and much less expensive than quality full-size automotive spray guns, and they are very handy for all the nooks and crannies that are difficult to spray with larger automotive guns. The drawback is their limited capacity. You will be constantly refilling it.

Another painting technology is high-volume, low-pressure painting, or HVLP for short. These sprayers have the advantage of generating less overspray, which is better for the environment and saves you paint. You can purchase HVLP units that run off standard compressors or as an entire system that generates its own air. Both types tend to be pricey. Since most folks restoring tractors end up needing an air compressor for other reasons, I recommend purchasing HVLP guns that run off a regular compressor.

Before painting, you must realize that the object of painting is to create a single homogenous layer of paint among all the layers of finish you apply, from primer to final coat. These layers,

if properly applied, create one single layer because they bond to each other through chemical reaction. Paint that is applied as distinct layers that rely on adhesion between coats is much weaker than paint that is applied as a total system where each coat "melts" the coat below it to form a chemical bond.

The first coating is primer, and preparing the metal surface for primer is very important. To prepare the metal, use phosphoric acid to etch the metal. Some folks use muriatic acid and while either will work well, phosphoric is more popular. Etching will create the rough surface needed for the paint to adhere. Some primers also etch the metal while they cure, and hence are called self-etching primers. These primers do not require a separate etching step. Before applying a finish, let me remind you to thoroughly clean the surface with a wax and grease remover. This will help adhesion and also help prevent a condition in the paint called fish eyes.

I recommend using a self-etching, epoxy-based primer. This stuff sticks like glue (it basically is glue), etches when it is applied, and is nonporous. It literally is the best primer by far for protecting sheet metal and adhering to the metal. The drawbacks are:

Convenience: The primer comes in two parts that must be mixed, usually in a 50-50 ratio, and then allowed to sit, or "digest," for about 15 to 20 minutes before you apply it.

Here we have body filler and two common paint choices: manufacturer and aftermarket. Whatever you use is up to you, but I would use the paint with the more correct shade. I would not let brand name drive my decision since most all national brand paints are good. As for body filler, I realize Bondo has endured sneers for ages, but its modern epoxy filler is as good as any on the market, at least for our uses. Just remember body filler is not a substitute for bodywork. Use proper metal-working skills to repair the parts, and use a sparing amount of body filler to cover tool marks and the small and slight imperfections left behind from proper body work procedures.

All of these compounds can be used as a thinner or reducer with paint, additives, or primers you might encounter. You should use the reducer recommended by your paint manufacturer. Be aware that they might try to sell you their name-brand reducer, but press them for the name of the common chemical their reducer contains and buy the less expensive generic form of the reducer. Here we have, left to right, acetone, lacquer thinner, Ditzler Paint's enamel reducer, and Toluene. Xylol and a few others are also commonly used. Remember that all of these are all dangerous chemicals and must be handled with care, gloves, and a respirator.

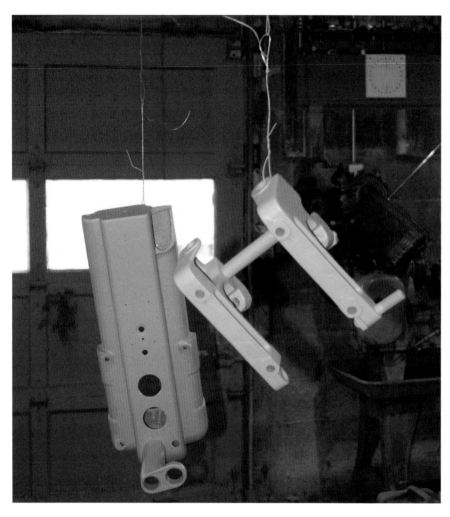

Painting irregular-shaped pieces usually requires that you paint them while the piece is hanging up. Here two pieces hang by a section of pipe from garage door rails.

accustomed to using. The second part of the two-part fillers is a hardener. If you use too little, the filler may take hours to cure and still be relatively soft. If you use too much, the filler will cure and become unusable right when you mix it. The mixing ratio is tough to judge and simply requires experience and testing. I mix a golf ball–sized portion of the filler with a marble-sized portion of hardener. Check the label first, but this proportion usually works well with most makes of filler. Finish up after the filler cures and hardens by sanding the filler smooth, making sure the filler conforms to the lines and curves of the metal. At this point, you should have sheet metal that feels very smooth with only very minor surface imperfections.

Now you will build up the primer coat with a type of primer called filling/sanding primer. The purpose of this primer is to build up the primer base, thereby creating a very smooth surface and filling the very small imperfections. This primer is meant to be sanded between coats. This levels high spots and smooths areas where the body filler may still be a bit rough. Several coats are needed, and professional painters usually alternate colors of this sanding primer, helping them to see during the sanding phase where their high spots and persistent imperfections are. I see the use of sanding and filling types of primers as optional with antique tractors. The fit and finish of antique tractors is loose and imperfect enough that using this

Price: Gallon for gallon it is more expensive than most primers.

Recoat window: This third drawback is related to its primary virtue—this stuff is so hard that you must apply your next coating to the primer within about seven days or the primer must be mechanically abraded and another coat of primer must be reapplied.

Other primers do well, and I do not specifically recommend against them, but in my experience I have

had the best luck with epoxy-based primers. Lacquer-based primers do very well also.

The next step after applying your primer is to apply all your body filler. This is where all your creases, minor dings, and deep scratches are repaired. Wrinkled areas left from repairs, welding pits and holes, and so on are now fixed with filler. The best body filler to use is the two-part lightweight fillers. These are polyester based and hold up better over the older, one-part fillers many of us are primer brings results much too modern and historically inaccurate to be considered necessary. A few coats used possibly to fill minor pits left behind by surface rust is all most people will need. In fact, I prefer to skip this step and use a special type of body filler called "glazing putty" to correct the very minor imperfections a sanding primer would be used to hide.

After priming you are ready for what is called the color or finish coat. All this work just to get to this point! There are two choices of coating you can use for the finish paint. Enamel is the most popular and is the coating that is historically accurate for most tractors built after the mid-1930s. Tractors made before that time were more than likely painted with a lacquer-based paint at the factory. If historical accuracy isn't important to you, I recommend enamel paint. It is easier to apply, less likely to cause problems down the road, available in a wide range of colors and custom color matching systems, and provides a finish that is every bit as beautiful as lacquer. Of course, there is nothing wrong with lacquer paint, and if you are comfortable using it or just simply wish to use it, then my instructions here will still be helpful and useful. My instructions assume, however, that you are using enamel.

Next, you will have to determine what types of additives you will have to add to the paint. The three most predominant are hardeners, reducers, and thinners. Hardeners increase the hardness of the coat after full cure and are an additive I recommend.

Hardeners also tend to increase the gloss of the paint and speed drying times. The next additive to consider is a reducer. This additive slows down the paint drying and curing process. This is handier during hot days or if you are using a hardener. Slowing down the drying process will tend to help the coat flow smoothly and help eliminate surface textures and other irregularities that detract from the finish. The third additive is a thinner, and this helps to improve the spray pattern from the gun, and improve the paint's ability to flow and become smooth on the surface. Depending on your circumstances and the quality of finish you are trying to achieve, you will need to use any combination of these three additives. Your paint supplier will be able to give you guidance with the particular products you are using.

Now you will spray two to five finishing coats, depending on what part of the tractor you are painting, amount of thinner and

When painting wheels, you should at least cover the insides of the rims with a nonporous primer, like epoxy, to protect against rust. These wheels have received a full coat of paint.

reducer, and quality of paint. I find two to three coats on cast iron is sufficient, and then I may use five to six coats on sheet metal. You should apply coats of paint in even, back-and-forth motions that cover the surface completely, but leave the paint thin enough that it doesn't run or sag. In short, several light coats are usually better than one thick coat, but if you leave the coat too light and thin, it will have a very grainy, dusty appearance that significantly reduces gloss and shine. The trick

Antique Tractor Tip

Applying Paint

If you apply coats of paint that are too thin, you will get a dry, dusty-looking coat of paint; too thick and the paint will run. The trick is finding the happy medium. Complicating matters is the fact that this happy medium will change between brands of paint, temperature, and so on. Be willing to be observant and flexible each time you paint.

This tractor shows the typical approach to painting an exhaust manifold. A high-temperature black or silver paint is used on this manifold that was sandblasted to remove all traces of rust.

too soon, the new paint will cause the last coat to wrinkle up. The other problem is to wait too long between coats. Then you have two coats that rely on adhesion instead of forming a chemical bond. Most modern enamels are forgiving and will allow you to wait up to two to four weeks between coats without adverse effect, but it is best to recoat as soon as the previous coat has dried to the touch. Most coatings will have some type of guidelines for recoating in the instructions or data sheets. Do realize these times will vary depending on temperature, humidity, and so on.

You will repeat adding coats until you have built up a good, thick layer of color that will withstand weather and time and the final finishing process that follows. If you have applied the paint right, the finish will be smooth, and for the most part, glossy. Much of the gloss will come from the final finishing, but the coating at this point should look smooth and bright already. This is the point at which you will apply a clear coat if you want to use it. It is historically inaccurate and unnecessary to clear coat a tractor, but many folks like the rich, wet-looking shine clear coat lends to a paint job. Typically, clear coat application begins with a mixture of clear coat and paint. After that coat, the clear coat is applied by itself, building it up until you have a good, rich shine and luster. Much of the same techniques for applying paint apply to clear coats. Additives are usually not added to clear coat.

to applying paint is practice, practice, and practice. Getting just the right amount of paint on the tractor takes judgment that is only learned from experience. Coats are applied as the previous coat has had a chance to dry to the touch. Remember, there are two main ways to create problems between coats. The first problem is not waiting long enough between coats. If you apply the next coat

Antique Tractor Tip

Ventilate with Care

While it is tempting to use ventilating fans while you paint, you have to be careful where they are blowing, and if the overspray is thick, you must use a fan with a combustion-proof motor (a motor with no exposed spark).

FINAL FINISHING AND DECALS

After painting, the part that brings out the gloss and shine begins. Whether you sprayed clear coat or not, the dust nibs, granularity in the finish, and other slight surface irregularities and faults are sanded out. This process is called color sanding. After the color sanding, you then buff and polish to enhance the luster of the paint and wax it to protect the paint. These procedures should only be done after the paint has fully cured, which can take several weeks for certain finishes, most notably enamel. To determine if your finish is ready, try to leave a mark in the finish with your fingernail. When the finish has cured enough to color sand and buff, your fingernail will not leave a mark. To color sand, evenly and completely sand the surface of the tractor using a 600-grit wet or dry sandpaper, being sure to use lots of water as a lubricant. Then repeat, moving up to finer papers in increments of 200 grit. For example, the next paper to use would be an 800 grit. Stop at 1200 grit. Then start buffing the paint with polishing compounds. I recommend Mequiar's polishing system. At that point, you can wax the finish. Congratulations! You should now have a very smooth, professional coating that will protect the tractor for your descendants, who will surely appreciate the extra effort you took to protect the tractor so they, too, could enjoy it.

After the restoration, the last thing to do is to apply decals. Most brand and model markings were originally silk-screened directly onto the metal at the factory. Today these markings are available as mylar decals. Here is a small sample of what you can expect when you purchase your decals, which are available through Internet store fronts and mail-order houses listed in hobbyist magazines.

Applying Decals

Applying the lettering to the tractor is straightforward, but like almost everything involving restoration, there are tips and tricks that will enhance the look of your work. Most decals purchased are made of mylar. That means the lettering is part of a clear plastic strip backed with adhesive. This clear plastic between the lettering is distracting and can create a poor look for your newly painted tractor. I suggest cutting out the letters and then applying them individually. You can use masking tape to mark the spacing of the letters before you cut them out. Other options for decals are having a local sign or graphics firm cut vinyl lettering for you. This can be expensive, if not impossible for multicolor decals, and you must provide original artwork that shows the shape and style of the letters for them to work from.

The last option, and the option that is the most original in many cases, is silk-screening the lettering. This was actually how the lettering was originally done at the factory. The problem here is that obtaining the silk screens, or having the work done, is very costly. Usually the only option is to ship the pieces of the tractor that are lettered to the few shops around the country that do this type of work. If you have a very rare or important tractor, I would advise this method. If a source of decals is not available and silk screening is out of the question, then vinyl lettering would be the method of choice. Otherwise, the mylar decals are your best bet. Do be careful: There are many decal sets out there that are inaccurate. Be sure of the authenticity and correctness of a decal set before application.

CHAPTER 17
Shop Equipment and Design

Restoring and storing antique tractors will eventually require some sort of special purpose building. The common solution is to build a shop, and this building fits the bill. It's spacious, has high ceilings, and uses a chimney for a heat source, which is a necessity in cold climates. It also has plenty of windows for letting in a lot of natural light, which is a real plus for working on and painting equipment.

Collecting antique tractors is one thing, but restoring antique tractors is quite another. If you plan on performing high-quality restorations as an active hobby or sideline business, get ready to spend some money. There are many procedures that are affordable to hire out to professionals when you have them done just once or twice every five years. But if you start performing these same procedures several times a year, it will usually behoove you to buy the tools and equipment necessary to do the work yourself. An example would be a small MIG (GMAW, or Gas Metal Arc Welding, is technically the correct name) welding machine. These machines are quite expensive, starting at several hundred U.S. dollars for the smallest machines. These machines are indispensable for welding sheet metal on tractors. Buying one isn't necessary if you just restore a tractor once every few years, but if you will have a lot of sheet metal repair that will need to be done, you should plan on finding training on this technology and buy the equipment to do it yourself. Of course, this is just one example. There are many other shop tools and machines that would make sense for you to buy if you plan on doing many restorations.

Air Compressors

One of the very first shop machines I recommend to anyone contemplating a restoration is an air compressor. This is the handiest thing since sliced bread and you will find a million and one uses for it. In addition, there are several shop procedures that will require an air compressor. If you ever plan on painting a tractor yourself without using spray cans of paint, you will need to purchase a paint gun. All paint guns run on compressed air. A sandblaster is another tool that requires an air compressor. There are several other wonderful tools that run on compressed air that make things much easier. Things like impact wrenches usually make short work of stubborn, rusty bolts. Air power scalers, grinders, and sanders also save on the elbow grease and save considerable time. A good air compressor should be your first purchase.

When buying an air compressor, first decide on the air tools you plan to use. Look at their cfm, or cubic foot per minute, rating (see below for a warning on tool cfm ratings) and note the rating of the tool with the highest air usage. Then buy a compressor that will supply all the air the tools need. If you believe that you may regularly use two tools at once (i.e., two people working together), be sure to add up the cfm ratings of the tools you plan to run simultaneously. Be aware that often the rating is the absolute minimum and the tool can and should be supplied

When adding a garage door to a new or existing shop, it should be as large as possible. Even if your tractor will fit in a smaller opening, you never know when you might want to back the tractor into the garage while it is on a trailer. The next tractor you purchase may be taller. Make the door as large as possible to head off future problems.

with more air than the rating. Something like a sandblaster becomes more effective with every cfm of air that you give it. By the way, you often will see the tools rated with a compressor horsepower rating. Do not use these horsepower ratings, as they are misleading and difficult to quantify if you use a couple of

Antique Tractor Tip

Old Air Compressors OK
Consider buying a used air compressor for your shop. Odds are the average antique tractor collector won't use it enough to push it to its demise, and if you do, the

compressor heads are easy to service and overhaul, and the motors are not horribly expensive to replace.

A block and tackle on a trolley track is a great way to solve the problem of lifting heavy loads, especially in low overhead buildings. While their lack of mobility can be an issue (you have to bring the load to the block and tackle), it isn't a frequent issue. You must be very sure your building can handle the loads this will place on your second floor or roof joists. Most buildings built with trusses can't handle large loads more than a few hundred pounds.

Antique Tractor Tip

Buy New Welders
Beware of used welding machines. Used MIG welders typically have defective wire-feeding mechanisms and stick welders often have transformers that are on the verge of failure.

tools at once. Cfm ratings are much more exact and reliable.

One warning on tool cfm ratings: Often they are based on a certain percentage duty cycle; that is, the manufacturer expects that during any given one-minute interval, you are only expected to use the tool a certain percentage of time. In other words, the manufacturer of a tool with a 50 percent duty cycle is expecting it to operate 30 seconds out of every minute. I usually find these duty cycles very misleading and unrealistic. If a tool says 50 percent, I double it to find the continuous-use rating. Then I look at the compressor rating I am planning to buy and see what type of duty cycle it will support. Here is an example: Let's say I have a die grinder that requires 5 cfm at 90 psi. The package also says a 50 percent duty cycle. I double it to find the continuous-use rating. This would be 10 cubic feet per minute. If my compressor, or the compressor I am considering, only delivers 7.5 cfm at 90 psi, the air in the tank will be expended every 45 seconds and I will need to stop and wait 15 seconds for the compressor to catch up. That translates to a 75 percent duty cycle. A similar calculation can be made for any duty cycle to determine how well your compressor can keep up with a tool.

Buying an air compressor is also simple. First, ignore everything the manufacturer includes in the product packaging, promotional labeling, and sales brochures. Forget all about any horsepower ratings. Instead, follow the cfm rating found on the specification plate located somewhere on the air compressor itself. Sometimes this figure will be expressed as cfm for standard cubic feet per minute, but for all intents and purposes, the measurement is the same. This information will be listed on the machine for two different pressure settings. Usually they are 40 and 90 psi, but sometimes the rating is given at 40 and 100 psi. The strategy is to buy the compressor with the largest 90 psi cfm rating you can afford, but always buy the minimum required by the tools you plan to use. Paint guns only require a few cfm at 40 psi to operate. On the other hand, die grinders, sandblasters, and some other types of tools will eat up every cfm you can give them. If sandblasting is a priority, there is no such thing as a compressor that is too big.

There are, of course, a couple of mitigating circumstances and other issues that need to be addressed. If you only have room for a portable compressor or simply want a portable compressor, your choices are limited by that criterion. If your price range includes a mixture of 220-volt units and 110-volt units, it is much better to buy the 220-volt unit if at all possible. This, of course, requires you to install a 220-volt outlet in your shop area. One other issue to resolve is whether you should buy an oilless versus oiled compressor. Stay away from oilless compressors. These compressors have nylon sleeves in the compressor block and therefore the compressor does not need oil to run. The nylon wears quickly, though, and does not have the longevity the oiled compressors

have. Definitely spend the extra money on an oiled compressor. Oilless compressors are intended for light-duty homeowner use. They are not intended for people who will use their compressors on a regular basis. One last thing: Buy the compressor with the biggest tank you can afford or fit into your shop. Compressors store air in the tanks, and if your tool requires more air than your compressor can directly provide, then a large air supply will be handy to prevent the delay caused by loss of pressure. The drawback to the larger tank is the compressor takes much longer to pressurize the tank.

Welding Machines

Anyone around antique tractor restoration will find that hiring out welding work is expensive, and often much welding needs to be done to tractors and implements when restoring them. It is only natural then that restorers tend to get training in welding, and then start buying their own machines. The first consideration is cost. A MIG welder will be important for sheet metal repair, but a MIG unit small enough to be affordable will not be big enough for repairing thicker metals. The alternative seems to be buying an AC/DC stick welder for the bigger, thicker metals and buying a small 110-volt MIG unit for sheet metal repair. This is marginally less expensive than buying a MIG welder capable of welding thicker metals. If you have a little extra money, buy a high-quality 220-volt MIG unit capable of welding anything from 22-gauge sheet metal all the way up to 1/2 inch.

This is an example of a natural gas shop heater and the chimney made for the unit. These units are a bit more involved and costly to set up when compared to a wood stove, but they heat the shop much quicker and with less work than splitting wood and firing stoves. In addition, in shop buildings that are often poorly insulated, gas heaters may be the only units capable of producing the BTUs necessary to heat the building.

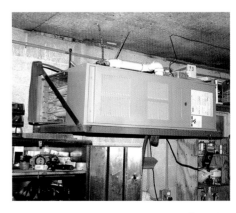

Unless you live in a very warm climate, you'll need to add heat to your shop if you want to work on your tractor year round. For most folks, a wood stove such as this works well. Other solutions include coal and oil stoves.

Another excellent choice for metal work is oxygen/acetylene welding equipment. In the hands of a trained professional, gas welding provides the smoothest, nicest-looking, and strongest welds possible. In fact, the FAA will only allow gas welding in the restoration of old airplanes that use lightweight structural steel tubing. You can use an oxygen/acetylene outfit for a variety of other purposes too. These gas welding rigs come complete with a lot of different tips, or ends, that will turn the torch into a heating tool, cutting tool, or welding tool. These are great for heating up metal parts so you can loosen them, shrinking sheet metal during sheet metal repair (by heating the sheet metal and then rapidly cooling it), and cutting metal during fabrication and repair. Welding with oxygen/acetylene can be slower than other methods, however, and most people find it more difficult to learn than the arc welding that is more commonly done these days.

Lifting Equipment

Any good home shop will incorporate some device for lifting engines, transmissions, and other heavy parts. Is it better to go with a block and tackle from a roof beam, or a portable hoist? The answer, as usual, depends on your circumstances. Portable hoists are expensive and require smooth concrete floors. Building hoists are even more expensive, but they don't require the smooth floor. Building hoists usually have limited travel, if they travel at all. Typically, people use portable hoists if they have concrete floors, building hoists if they don't have them. One last related item is an engine stand. If you plan on a complete remanufacture of your antique tractor's engine, these are a must to hold the block while you assemble the engine. As a side note, many modern engine stands do not have support plate bolt holes that line up well with antique tractor engines, so be prepared to make adapter plates, or have them made.

Other Equipment

There are many other shop tools and machines that you may want to consider buying. A valve-grinding machine can be handy, as can a metal lathe and a metal mill. Two tools that will make the process of prepping all the tractor's small parts for paint much easier and quicker are a parts washer and a small sandblasting cabinet. Some other tools you'll find handy are a bench grinder, electric hacksaw, and heavy-duty drill press. These all are time-saving devices that will help get a job done quickly and well. As you continue to expand the types of projects you undertake, you will discover new needs and new tools to fill them.

Antique Tractor Tip

Level-Ground Shop
While pole buildings make excellent shops, their use on anything but level ground is restricted. Otherwise you will spend more money grading the building site to achieve a level spot than you would spend on a traditional masonry foundation.

Designing a Shop

All my life I have always worked on cars, trucks, and tractors under shade trees, drive-in basements, and sheds. I have never owned a very nice shop and don't want to give the impression to anyone that it is an absolute requirement. There is a price to pay for not using a shop, though. Makeshift shops such as I have often cannot be used during certain types of weather, force you to perform extra work, and in the case of drive-in basements (my current shop), possibly expose the house and household to the risk of fire. This risk forces you to do certain things outside, such as painting (it is unhealthy to paint in a basement where the household, and you, would be exposed to trapped fumes). Painting outside has its drawbacks, as when an unexpected breeze picks up and blows all types of dust and grit into your work. The sanding and buffing required to remove these impurities can take several hours. Therefore, I think you'll find that decent shop space is a very important priority if you plan to do a lot of restorations, plan to restore large tractors, or simply want to set aside a work area that is yours and yours alone. Well-designed shops are also safer and tend to help keep the clutter out of the yard and yield more professional results.

While I have never owned my own shop building, I have worked in many different shops and have had the benefit of being associated, through my Internet site that is devoted to antique tractors, with a survey of antique tractor restorers regarding shops. The survey was a general one about their shops: what they like about them, what they would change, and what their dream shop would contain. The information contained here is based on my own experience and the results of the survey.

First, the majority are pole built. That means they are designed using post and beam framing, but important structural posts, or poles, are sunk into the ground. Most of these structures seem to be metal sided since this is less expensive than wood siding. Roofs are almost always metal and usually insulated to prevent condensation. The floors are at least packed gravel, but this usually seems to be a stop-gap measure until the owner can afford to pour a concrete floor. Personnel doors are at a minimum 3 feet wide (this is usually required by code, anyway) and equipment doors are 10 feet wide by at least 10 feet high, or if they are smaller, the owner wishes they were at least this big. Windows are a personal preference, but skylights seem to be a nice extra folks like to add to increase natural light (important while painting). This seems to fall into the "if/when I can afford it" category.

It is common for almost all shop owners to say that their shop is too small, and if they had it to do over, they would make it bigger. In fact, this is so common and universal a sentiment, I would advise you to build the biggest shop you can

I have found that having a full selection of cotter pins, grease fittings, O rings, and other miscellaneous hardware is a necessity in the shop. There is nothing worse than having to stop a repair or restoration because you were missing the right size of a cotter pin. Fortunately, these parts are available as collections in boxes that contain a myriad of sizes. Be sure to have these on hand in your shop.

Many parts on an antique tractor are pressed together, which means they are held together by the fact they are about the same size and stay together by friction. An example would be the shaft and bearing assembly in a water pump. Disassembling these press-fit parts requires significant force. The only tool that will press them apart is an arbor press. This is a small bench-top unit for smaller parts. Large floor units for larger parts are affordable as well.

As you set up a work area to use as your shop, putting together a workbench should be your top priority. This workbench illustrates how it should be done. It's well lit and has plenty of storage and outlets, a generous work surface, and a couple of nice vises.

afford. It will fill up fast. My work areas have always been so cramped I never would have guessed that some of these sizes would seem small to others! Some common dimensions are 24x24, 30x40, 30x80, 42x48, 24x48, 30x60, 36x36, 25x25, 30x64 and 40x60. Ten-foot and 12-foot interior heights are used most often. Plan your shop's dimensions in multiples of 3, 8, and 10 feet to save on building materials. Also consider rough green wood from local mills to save money.

Electricity requirements seem to be a standard 200-amp service panel. Shop machines such as air compressors and welders use up tremendous amounts of electricity, and a 100-amp panel will not be sufficient. If these large machines are not in your future, 100 amps may be adequate. Phones and water seem to be a matter of personal preference. Some don't like the distraction of the phone in the shop while others hate to walk to the house whenever they need to use the phone. General safety preparedness dictates having a phone available for emergencies, if possible. Bathrooms are not common, and constructing the septic field is the main expense that discourages them. Hot water and a hand-washing sink are the main motivations for adding water to a shop. Some people even add cable television to avoid missing important shows and ball games!

Designing each work area in the shop is very particular to each individual, but there seem to be some common themes among all the restorers. The first is, most shop owners would love to have an old-fashioned oil change pit but were amazed to find out they are dangerous and most insurance companies will not insure a shop with one. The main reason is that most volatile and combustible fumes are heavier than air and any gas spilled in the pit will turn it into a flaming death trap when a spark is caused by a dropped tool. Most restorers' shops have, or restorers wish they had, a separate partitioned area for a "clean" room. This is where painting would take place and would be an area easily cleaned and made dust-free, even if this means using moveable partitions. Another area many either like or wish they had was an area that could double as a dirt room, where tractors and other equipment could be hosed off and sandblasted and the dust and water would stay put and out of the rest of the shop. This area should have adequate ventilation, so plan for this if you decide to build this type of room. Most wish there were a

Antique Tractor Tip

Don't Forget About Light
Lots of natural light is an advantage in a shop. Be sure to design your shop with a lot of windows.

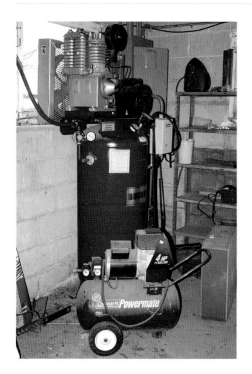

Your shop compressor should be fairly large with lots of air storage, but a small portable unit, seen here at the foot of the large compressor, is great for generating air away from the shop.

When using an air compressor for an extended length of time, the air coming from the compressor will be hot and the water vapor in the air being supplied will condense in your air supply pipes, hoses, air tools, or paint if you are using a paint gun. This is a shop-made air cooler designed to cool the air being supplied and allow the water vapor to condense in this large copper pipe. A valve for draining the accumulated water is at the bottom of the compressor.

separate heated area with chairs, a desk, and table to take a break in during the winter when the rest of the shop is cold. One last thing that was mentioned a few times by respondents was having a tractor trailer dock. If you ever have to transport large tractors, building a loading dock may be a smart idea.

Heating a shop takes on many different forms across the country and reflects the regional variations in heating costs of various types of fuel. Wood-fired stoves and water stoves are common in areas of the country such as my home state of North Carolina where wood is plentiful. But in the coal-producing states you find coal stoves and LP gas furnaces more common. Whatever works best for you is the most important

thing. There are some safety considerations: stoves and other open sources of ignition do not mix with paint fumes and other flammable liquids associated with antique tractor restoration. Be sure you consider safety when designing shops.

Just a short word about the construction process. In many agricultural areas, obtaining a permit for an agricultural building is not necessary. This is true in my town, but I do know for a fact that in many communities it is necessary to obtain permits even for agricultural buildings. On top of that, while we consider antique tractor restoration an agricultural activity, many communities do not, citing the fact that it is no different from any other collecting activity. In my community, the building inspector said that as long as the owner used one of the tractors, a building used for antique restoration would not require a permit. If, however, the owner did not use any of the tractors, and simply kept them as a collection, then the building would be considered personal shop space, and a permit would be required and the building would be subjected to other zoning issues and inspection procedures. Utility companies present similar issues. Often many utilities, electrical power utilities in particular, have different rates for their service and product if the building is considered shop space or agricultural space. Be sure they are aware of your intended use when they connect the building.

Shops and shop machines add to the enjoyment of restoring

antique tractors. If you have ever had to hand-drill through a hardened steel shaft while under a tarp in a cold rain in the winter, you will understand what I mean. But the expense is not insignificant and building one requires time and thought. In fact, just like tractors, how much you get out of the exercise of building a shop and acquiring machines will depend on the time, energy, money, and thought you invest. Just be sure safety is a number-one priority and the decision you make reflects your real needs and desires.

CHAPTER 18
Machining and Metal Fabrication

When metal parts have to be repaired beyond the minor grinding and welding most of us can handle, or new parts have to be made, you'll need the services of a machine shop. However, George Willer of Ohio has the skills to do his own work and has the proper machines in his own shop. In the foreground is a milling machine that can perform a wide variety of metal working tasks, including milling a surface flat, cutting grooves, and more. To the left is a lathe for machining cylindrical parts, and behind the mill is a metal saw. He also has a full complement of smaller metal working machines and tools.

The fabrication and repair of parts, accessories, customizations, and implements requires another set of skills that many antique tractor restorers find worth acquiring. Metal work, like any other set of skills, requires patience, proper instruction, and a certain minimum set of tools to get started. The biggest advantage of becoming proficient in metal working is the ability to continue a restoration that would otherwise be canceled or become prohibitively expensive because the restorer was unable to make or repair an expensive part himself. Invariably the same pieces of sheet metal are missing or irreparable on almost every example of a particular model of tractor. For example, battery boxes for Farmall tractors are always rusted out. Finding a used battery box in good condition is difficult. While a reproduction item is available in this example, they are not available for all parts on all models and may be in short supply even if they do exist. If you have the ability to craft your own sheet metal parts, you gain flexibility in determining the model and time frame for your restoration project. Making the parts may also be cost effective. Simple pieces of sheet metal, brackets, and many other parts are within the abilities of the average restorer. A local foundry may be able to provide you with the cast iron and steel pieces that you might need if you can create a prototype out of resin for the mold. Ingenuity usually triumphs in situations like this, so don't let a lack of parts be the reason your project is postponed or canceled.

While making your own parts is sometimes necessary, antique tractor owners often end up doing their own fabrication work to accessorize and customize their tractors and implements. Beginners and novices to metal fabrication should not be making structural and safety-related items for their tractor, but there is much else that can be done by the novice. Items commonly made by owners include a step to help get on and off the tractor, an extra toolbox, weight boxes for the tops of implements, and weight brackets for tractors used in competitive pulling. You may think of many other parts it would be convenient to make and add. Welding can be dangerous, though, so be sure you learn and follow safe welding procedures. I suggest a metal fabrication course at your local community college to help you get started in metal fabrication and welding.

Some shop-built implements are less about work and more about play. Here is a nice ice cream churn powered by the tractor and carried with the three-point lift.

Sawing and Drilling Metal

Metal, unlike wood, requires great patience and slow speed to cut and drill. This is especially true if the metal has been hardened.

Metal work also requires greater precision and tolerances than one might need for a project made of wood. For these reasons, small, portable electric saws and drills are of limited use. A floor-standing drill press and a motorized hacksaw should be considered if you plan on doing much work with structural dimension steel, such as channel bars, rods, angle steel, and so on. These will help you finish faster and with greater precision than you could achieve with hand-held electric tools. When sawing and drilling, use quality bits and blades and cutting oil, an oil designed to reduce friction and to cool the blade and the object being worked.

If you are working sheet metal or other thin-gauge metals, you will find that cutting the metal is fairly easy, and it can be done with a jigsaw equipped with a metal cutting blade, nibblers, or shears. The latter two are most often air powered, working much like a pair of scissors. They leave a cleaner and smoother edge than a jigsaw and are faster (but not necessarily more accurate). Nibblers and shears are by no means tools you must invest in, but if you find yourself doing much of this work, you may want to consider them. Shaping the metal requires a good selection of hammers, dollies, and anvils. Other shop items for those getting into extensive sheet metal work are brakes and rollers. These items are designed to help you form curves, angles, lips, and so on.

Heat Treating

Occasionally, after fabrication, the metal has to be made hard to

You will need to drill a hole in a hard steel part in a precise and careful way sometime during the restoration process. This is impossible with a hand drill and requires a drill press. Here we see a mid-standing (not really a bench-top, but not really a floor standing) drill press that will serve all your needs.

withstand the use and rigor of field work. A perfect example is a drawbar. A mild steel drawbar will quickly bend, buckle, or at least develop elongated pin holes if the steel isn't hardened. There are professional heat treaters that can take care of this work for you. Heat treating hardens the metal by causing the crystal structure of the steel to tighten up. This occurs when the steel is heated to a certain temperature range and then rapidly cooled, traditionally in water or oil. Because of the rapid cooling, the crystals become smaller and more numerous, causing greater density and stronger bonds between

the crystals. This is one aspect of metal working you shouldn't bother trying to handle yourself unless the part is really small. Proper heat treating is tricky to do right, and experience is the only teacher. A small part can be hardened by thoroughly heating it to a cherry red and then quenching it in water. Until you are proficient and trained, do not quench in oil since the risk of fire is present. Varying degrees of hardness can be obtained by changing the temperature of the item before quenching and differing the rate of cooling.

Restoring a tractor involves replacing worn-out parts, but often it makes better sense to refurbish the part, bringing tolerances back into specifications. Examples are pins that have developed a lot of play within the housings, and bushings. Another common example of a highly worn area is the throttle. Throttle shafts and indexing plates often have teeth that are worn, yet the parts are too expensive or too commonly worn to warrant replacement. Whatever the part or the reason for wear, repair of the worn areas should be an option in your mind before replacement of the parts. Sometimes, as is usually the case in throttle assemblies, the areas in question are welded up using a low-heat welder such as a MIG welder turned down low. Then the areas are shaped and finished with files and die grinders. Large pins and pivot spindles that turn in housings and bushings are repaired by replacing the

bushings. Sometimes turning a new pin will work if the wear in the housing or bushing is very even. If the pin or spindle was turning in an unbushed housing, and the wear in the housing is not even, sometimes drilling out the hole of the housings to a larger size, and then using a bushing, is the proper repair. Some careful study and ingenuity will help you repair highly worn areas of your tractor that don't lend themselves to replacement.

Press fit parts are parts that stay in place on the tractor because they are driven into housings or onto shafts that are sized for a tight fit. That means the parts will stay put without fasteners. Unfortunately, they usually require significant force to remove. Performing this work requires the use of good tools, patience, and if possible, a manual or hydraulic arbor press. The most common example of press fit parts are bearings and their races. They can be removed with tools called bearing splitters, pullers, and drifts. Any combination of these tools might be needed depending on the circumstance.

The most dependable way to remove press fit parts manually is to generate some type of wedging action. Bearing splitters work on this principle. Drifts and pullers work well where a wedging action can't be generated. Slide-weight pullers can be especially effective. Sometimes even going to the extreme of designing and building your own special one-off tool is not as crazy as it might seem. Since most press fit parts can be damaged with hard steel tools, the tools are best made out of mild steel or even brass. These metals are readily worked with typical shop tools and saws. And coming up with a tool that is just right can mean the difference between success and failure.

Often parts with a symmetrical circumference will need to be returned to true, and there is no substitute for machining the parts on a metal lathe. Common examples of parts that usually have to be returned to true on a lathe are brake drums and plates, flywheel faces, electric motor commutators, engine shafts, and other miscellaneous shafts. Of course, there are many other parts you may run across that need lathe work. This type of work should be done by a competent machinist. Metal lathes are typically expensive and require training to use, but if you have one available you will find many good uses for it during your tractor restoration once you know how to use it.

Another common machine shop chore is the straightening of parts. Many parts of a tractor are subject to great forces during use, causing damage. Damage can also occur from abuse and well-meaning but ill-conceived owner modifications. Straightening the parts from these mishaps can often require forces greater than many folks can generate in their home shop. More often than not, these straightening procedures are best done when the parts are heated extensively. Probably the most common example of this type of repair is the straightening of steering tie rods. Not only do the tie rods need to be straightened, but often these rods are adjustable and are made up of two halves. Over the years the two halves have fused together. To straighten and loosen the halves, the shops will use a heating tip on an oxygen acetylene torch and heat the rod as completely as possible and then use a straightening V block and a hydraulic ram to return it to true. The act of heating and cooling will often loosen the rod halves and allow the rod to once again become adjustable. Likewise, I have straightened steering shafts, gearshift levers, frame rails, and other parts using this method.

Adding shop-made improvements to tractors is a time-honored tradition on farms. Here a small step has been fabricated by the owner to ease the effort required to climb onto the tractor.

A restoration of an antique tractor doesn't necessarily mean a strict adherence to originality. This Farmall Cub was upgraded with a set of shop-made dual rear wheels and dual seats. This great little tractor was restored and built by George Willer.

This is a clever shop-made system that provides feedback from the implement being used to the hydraulic control valve that provides the up-and-down control to the implement. Like most ingenious designs, it draws on simplicity for its strength. A bracket is attached to the three-point hitch side arm, which then clamps to a rod attached to the control valve. The clamping point is set a bit differently for each implement. The unit operates by pushing down (lowering the implement) when the implement raises up and pulls up on the control valve (raises the implement) as the implement drops. This system smoothes out the inconsistencies seen in implement depth and travel. This system was designed and built by George Willer who reports it is especially handy when using a scrape blade.

Welding

Welding is a skill that all antique tractor restorers end up learning or wishing they had learned some time or another. Before I go any further, I beg you to get training before attempting to weld any pieces of the tractor that are structural or could compromise your safety should the weld fail. With the help of an experienced friend, you can train yourself to weld noncritical components. These old machines often come to us with many flaws and repairs that can only be corrected and implemented using welding techniques. For our purposes, there are only three different types of welding techniques (I am using the common name of the technique, not the exact name). The three types of welding are gas, arc, and MIG. There are others such as TIG welding, whose application is uncommon for most antique tractor welding projects.

Gas welding involves using a gas-oxygen mixture with very high burn temperatures, typically in excess of 2,000 degrees Fahrenheit. The gas normally used is called acetylene, though there are others, such as MAPP (which is available in small torch types of canisters at your hardware store). Using the proper torch tip, the two pieces to be welded are clamped together, and then the pieces are in essence melted together using a small steel rod or other source of mild steel as a filler material (coat hangers work in a pinch). Both arc and MIG welding harness the power of electricity to "melt" or weld two pieces of metal together. Arc welding uses a filler rod, energized with electricity, to direct the current and spark to the joint of the pieces being welded. As the weld progresses, the rod is consumed and becomes part of the weld. MIG welding does much the same thing, but a MIG welder feeds wire via rollers and a small gun type of tip into the joint. This wire is energized by electricity, and as the wire is fed to the joint, the current melts the parent metals and wire together.

The main tricks to welding are properly preparing the parent pieces, using the right rod or wire for the materials being welded, setting the welding machines or regulators to deliver the right amount of heat, obtaining the proper amount of cut, and proper heat-up and cool-down. Properly preparing the parent pieces means removing all rust and dirt and getting the pieces lined up, square, and clamped. Some types of metals, such as cast iron, and thick pieces, should be preheated before beginning welding. Here it really helps to have an assistant available with a heating tip on an acetylene torch to get the piece hot and keep it hot while you weld it. Selecting the right filler material is very important also. If you are about to weld cast iron, you must use stainless steel or nickel rods and wire. Proper cut into the material means that you have sufficiently cut into the parent materials deeply enough to get a good strong joint, but not so deeply that you then compromise that joint. Overheating also may warp or change the geometry of the pieces. At the other end of the spectrum, not heating and melting the parent metals well enough results in a cold joint that is very weak. During the cool-down period, the weld should be peened with a chipping hammer to relieve the internal stresses that develop during cooling.

Choosing between the methods involves some tradeoff. MIG welding would be the technique of choice if the equipment were cheaper. Arc is great but you can't weld sheet metal with it (at least nonprofessionals can't, though I know some professionals who claim they can). Gas is not as fast and a bit tricky to learn, but oxygen-acetylene torches are handy around the shop for cutting off stubborn bolts, making a quick cut in steel, heating parts to loosen them, and so on. My recommendation is to eventually acquire all three types of welding equipment as you need them and can justify them. The gas torch is so handy, you should probably acquire it first. Most get a small 110-volt MIG unit after the gas torch. That is because it can weld sheet metal, something most novices can learn to weld fairly easily. After that they acquire an AC/DC arc welding machine. Welding is an enjoyable science, craft, and art form and as you become more proficient in antique tractor restoration, you will find more and more instances where you can put these newfound skills to use.

Metal fabrication and repair is within most everyone's grasp. If you

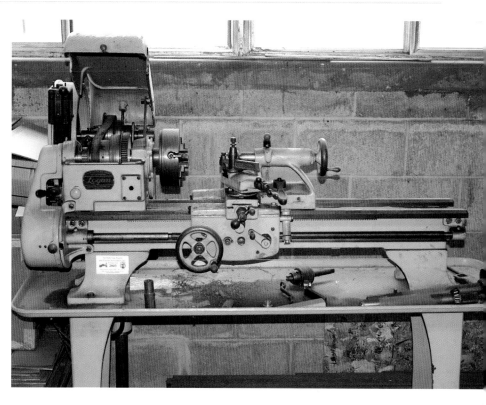

Owning a metal working lathe such as this might be dismissed by some as beyond the scope and ability of most antique tractor enthusiasts. Small machines typically are not horrendously expensive, and an old but solid and reliable unit can be purchased at many large antique tractor swap meets for a price similar to a good air compressor. Training on how to use the lathe is as near as your local vocational college. Anyone who has a lathe or has access to one and knows how to use it can save a lot of money and time compared to hiring a local fabrication shop to create the part or repair.

are mechanically inclined enough to work on an antique tractor, then you should be able to learn to make, repair, and weld your tractor's parts after a little training and practice. Just remember that most machine shop procedures are dangerous activities and safety is more important than anything else. Be sure to invest in all necessary protection such as safety goggles, welding lenses and helmet, gloves, fire extinguishers, and so on. As for eyewear, I recommend using wraparound OSHA-rated lenses for general shop use. During arc and MIG welding, I recommend lenses that are rated as a number

10. Some folks prefer a 9, which shields less light, and if you are comfortable with that lens, feel free to use it. Do try a 10 first, though. Gas welding doesn't require quite the filtering and UV-light-blocking properties required for the electrically based machines, and those goggles are much lighter and easier to see through. Don't forget to protect your hands, arms, and head as well as your eyes. You will also be amazed at how many times you will forget that a particular piece is still hot and pick it up. Don't take off your gloves until you leave the shop and are done. Have fun and be safe!

Shows, Demonstrations, and Pulls

The crowds are getting bigger at classic and antique garden tractor sales. Their smaller size and (typically) smaller costs than full-size tractors make them an attractive alternative for many antique equipment buffs.

There are many reasons to become involved with antique tractors on more than an occasional basis and there are many different ways you can enjoy them outside of your fields and farm. Restoring, showing, and competitively pulling and plowing with antique tractors is a growing hobby, with millions of people a year attending shows and competitions across the world, particularly in America. There is a niche for everyone who wants to share in the antique tractor hobby, whether you only want to drive a tractor in your local parade, or to participate in shows or competitive pulling events around the country. Simply making some contacts and friends who also share the hobby will help you find a way to turn it into a pastime.

Collecting

Collecting antique tractors, for the same reasons others collect stamps and coins, is a popular and growing hobby. Within 10 miles of my house are four collectors who have over 20 tractors each, and two of them have over 40 tractors! While one of these collectors uses some of his tractors for work around his farm, the other simply enjoys taking them to shows and participating in local pulls and parades. One parks all of his tractors beside the road a few times a year for others to stop and admire. Whatever your motivations and desires, collecting can be enjoyable if you have the space and money. Antique tractors, as a whole, are slowly appreciating in value, and it is commonly accepted within the antique tractor hobby that they will continue to do so for a number of years to come. While buying antique tractors as an investment is not wise, it is nice to know that you can collect something that for a while at least, will not lose money.

Some people collect on themes, and try to collect certain types of tractors. Maybe they collect nothing but John Deere tractors, while a friend may only collect Ferguson tractors because his father is a Massey-Ferguson dealer. I have one acquaintance who does nothing but acquire antique tractors that are in excellent original condition. To him the makes and models don't matter; he just wants to find tractors that are examples of how they looked originally from the showroom. He will point out to you all the runs in the paint, the crooked decals, and the other little nuances the restorers destroy when they repaint and restore their tractors. Whatever the reason, theme collecting can make the activity enjoyable and provide a sensible collecting strategy.

John Deere New Generation tractors, the four- and six-cylinder tractors that followed the historic two-cylinder tractors, have become quite collectible. This John Deere 4020 diesel will work the fields or make a nice addition to any collection.

Showing Antique Farm Equipment

Probably one of the most popular antique tractor pastimes is exhibiting equipment at shows. One reason this is so popular is because most of these shows welcome all makes of tractors in all conditions. Antique farm equipment shows, with a few exceptions, do not cater specifically to restorers with like-new tractors. The more the merrier seems to be the prevailing attitude. If you have restored your tractor, showing off your prized possession and the work and time you put into bringing it back to life is enjoyable. Trading stories and notes with other owners and restorers is another highlight of most shows and I suspect camaraderie is the biggest attendance motivation for most collectors. At these shows you can swap war stories about the

parts that were hard to find, the mistakes you made, and successes and triumphs along the way. Seeing real examples of all the different types of tractors is the best way to learn about them. Shows usually have parades, tests of operator skill, and pulls that you can participate in, giving you a chance to show off your tractor. If you want to turn collecting antique tractors into a hobby, attending shows is probably the first thing you should consider to begin developing and cultivating your new pastime.

Old-Time Farming Demonstrations

Often folks buy or inherit antique tractors that come with antique equipment that the farm tractor may have powered—perhaps a thresher or something smaller, such as an old cut-off saw, sorghum mill, rock crusher, and

Antique tractor pulling is divided into two main classes: stock and modified. Modified tractor pulling may involve antique tractor frames, but they are so heavily modified they barely resemble the donor tractor. Stock pulling requires the antique tractor to be in stock condition and the only modifications allowed are additional weights and safety improvements. Here a John Deere adds weights to the front to prevent the front from raising up during the pull, increasing the load on and traction at the rear tires.

so on. Many folks not only restore the tractors, but they also restore these additional machines and then operate them at shows to provide a live exhibit of how these farm chores were performed in the past. Many tractor manufacturers made the engines that powered their tractors available as separate power units. Some folks repair and restore these power units and take them to shows for demonstrations. These live demonstrations are very popular with the attendees of the show, and owners of antique

Antique Tractor Tip

Wide Wheel Adjustment
Set your rear wheels as far out as possible to prevent the sled from moving over loose dirt that your tires create during the pull.

tractors and other equipment enjoy putting on re-enactment demonstrations of old-time farming methods.

Competitive Plowing

I have never participated in a competitive plowing match nor do I claim to be able to open up a good-looking furrow, but some folks do and can. They really enjoy participating and competing in antique tractor plowing matches. In these matches, judges watch and evaluate the antique tractor operator's ability to cleanly, smoothly, and properly open a furrow in a manner that is effective and efficient. Things like furrow straightness, furrow depth, the plow's ability to scour and cleanly slice as well as the operator ground speed and attention to the machinery are all judged in unison to determine the winner of the match. Often plowing grounds are chosen for these matches that are highly variable in terms of topography, which forces the operator to plow more than a straight line. It is interesting to note that most plowing match participants do not use three-point hitch plows, as they do not create furrows as neat and as consistent as trailing plows.

Competitive Antique Tractor Pulling

Perhaps the most common, and for some the most enjoyable, way to show off an antique tractor is to partake in antique tractor-pulling matches. You can participate in

pulling without becoming highly involved. Any antique tractor will do the trick; the only thing you need is a desire to learn and have fun. Winning as a primary goal will leave you disappointed unless you immerse yourself in the sport and give yourself a reasonable, consistent chance to win. Pulling for the sake of enjoying and using our antique tractors is the motivation for those of us who don't care to take up antique tractor pulling as a hobby in and of itself. I like to attend shows and will compete in any tractor pull that the show hosts, but I typically do not attend an event that is strictly a tractor pull, nor do I ever compete for points in any of the pulling circuits. This middle-of-the-road approach allows me to learn and use my tractor more, teaches me to become more proficient, and allows me to spend time with my friends. For many collectors, this may be an opportunity to really use their tractor, and I encourage as many people as possible who are comfortable with trying to learn how to pull.

There are strategies involved with pulling and all sorts of things you can do to maximize the performance of your tractor. There are different styles of tractor pulls (percentage pulls, transfer sled pulls, and others), but this advice is pretty standard for all of them. You'll learn more about pulling through doing it yourself or talking with others who do it, but these tips should help you get started. Before you try pulling a heavy load, be aware that rearward flips are common on the pulling track.

I have seen pictures of tractors upside down sitting on a pulling track (no one was hurt in any of them, fortunately). While nine times out of ten this happens in the modified class where the owners make significant performance enhancements, it can happen with any tractors pulling a nearly dead load like a weight sled. Be sure you are trained and able to handle your tractor should the front end start coming up during the pull. I also recommend wheelie bars for your tractor to prevent the tractor from flipping rearward.

The trick to tractor pulling is to put as much torque on the ground as possible. Notice that I didn't say horsepower. Horsepower has a speed and time component that doesn't come into play during a tractor pull. While highly modified tractors use speed to their advantage, I am specifically referring to a stock antique tractor pull. In the stock pulls no tractor is rewarded for pulling quickly, so generating as much torque to the wheels as possible is the idea. The first problem you will run across with that strategy, though, is your drive wheels will slip if too much torque is applied. Overcoming wheel slip is best accomplished through deft throttle handling, low drive wheel inflation pressures, and added weight on the drive wheels. If you have consistent problems with wheel slip, even after making these types of adjustments, then weighting your tractor more and moving up to the next weight class may be worth a try.

The next tip is: Tractor pulling is a lot like wrestling. You should get down into the lowest weight class you can, be the heaviest in that weight class, and then train to be the strongest and smartest in that class. As far as being strongest in the class, be sure your tractor is running at the highest no-load rpm allowed for your class. Most pulls allow up to 10 percent over factory specifications, while some require factory specifications. Whatever the rules, follow them, but be sure to get as many rpm as possible. If you are competing in a modified class, there are many other things you can do to make your tractor stronger. If you are competing in a stock tractor pull, then your options are limited to making sure you have a fresh tune-up, gas, and so on after checking your no-load rpms.

The last set of tips concerns hitching up and navigating the track. Hitching strategies are designed to raise the front of the sled as much as possible to minimize sled-to-ground friction and to transfer the resistance of the sled to the drive wheels in such a way that wheel slip is minimized. Without exception, using as high a

Another solution to the problem of adding weight for an antique tractor pull is a steel drum affixed to the wheel to hold sets of factory doughnut wheel weights. Weights are easily slid on and off on this type of weight rack.

drawbar setting as you are allowed is best for pulling efficiency. You should also hitch as far from the rear axle as possible. This helps with traction earlier in the pull and will ensure that the weight of the tractor is on the rear wheels. The trick is to find a hitch length that lifts the front tires just barely off the ground. After that, you actually lessen the downward forces on the rear tires and risk rearward flip. When navigating the track, seek areas that are domed or ridged that will minimize sled friction. Look out also for areas of soft dirt. You

Antique Tractor Tip

Go to the Back of the Class
When deciding on what weight class to pull in, a good place to start is to register for the lowest weight class possible, then weigh down your tractor to be the heaviest in that class. Occasionally some tractors will do better in a higher weight class, but start in the lowest and move up if need be.

The Oliver 990 with the GM diesel and the Massey Ferguson 98 with the GM diesel can be seen battling it out on pulling tracks across the country in the higher weight classes. This Oliver is more a show and parade machine, as you can tell from the tires. A pulling tractor would have shaved tires with a different cleat design.

getting off to a good start is half the battle. Typically, you will not need full throttle to start out and in fact you will usually only need full throttle toward the end of the pull. The larger classes should concentrate on weighting strategies and hitch issues. The rear tire inflation pressures can be important and should be low, around 12 to 15 psi. Some use even lower inflation pressures.

Antique tractors owned and used as a pastime and hobby are common across the world. Some enjoy them simply as reminders of the past. Some enjoy them for their design and workmanship. Some enjoy the challenge of using them in competitive situations or simply because they like tinkering with and repairing old machinery. There are as many reasons for enjoying them as there are individuals. One of the most compelling parts of the antique tractor community is a near universal tolerance of all these reasons. In short, everyone feels free to enjoy and use antique tractors on his or her own terms. There are no standards and rules pressuring a person to collect and use tractors in a certain way. At shows, field rough tractors are usually welcome beside the restored tractors. There is almost never any kind of "restoration" judging, and participation in the competitive activities is always strictly voluntary. Feel free to bring your antique tractor to any show, regardless of condition, and come to these get-togethers ready to enjoy yourself and the company of others.

will lose traction while creating additional loads on your tractor.

Another thing you can do to be competitive enough in tractor pulls to have fun and really test your tractor is to learn how to work those brakes! The rear drive wheels will lose traction differentially—that is, one wheel will start slipping before the other. By applying the brake to the rear wheel that slips, you will transfer power to the wheel that is not slipping. Adjust your rear wheels so they extend past the edges of the sled skid plate. That way the sled is not running over the soft dirt your rear tires have a tendency to dig up and mound. In the lighter classes,

Antique Tractor Tip

Pulling Track Tactics
When negotiating the pulling track, avoid soft areas and try to get the sled on areas that have a slight "crown" to them. This will minimize the resistance the sled is offering.

APPENDIX

Old Tractor Almanac

Reference Information

While finding additional information about restoring tractors is not tremendously difficult, there is no flood of information available either. Finding most of it involves subscribing to the various periodicals available and finding books and other printed materials that pertain to antique tractors. There are several good periodicals that are devoted to antique tractor and engine collecting and restoration. Voyageur Press, the publisher of this book, also has many, many good titles on antique tractors. You can find older out-of-print books and periodicals at antique tractor shows and swap meets, and antique book and magazine dealers. Books like the series John Deere used to publish called *The Operation, Care, and Repair of Farm Machinery* are of great assistance, if you can find them. Through perseverance you can find the printed materials necessary to pick up many of the details, skills, and ideas you need to perform a historically accurate restoration.

Fortunately, printed material is not, by any stretch of the imagination, the only resource. There are many other avenues available and each provides information in its own unique way. Often simple things like asking friends and neighbors about any antique tractors or any past history with farming will yield tremendous insights and information. Folks that you are sure have no farming or rural background will surprise you with a past you never guessed. Agricultural fairs and antique tractor shows are other obvious places to look.

The equipment on display is just a small part of what's available to learn. Stop and talk with everyone you can at shows and fairs. You'll be amazed at the stories and experiences available for the asking.

Finding a Mentor

Another very valuable way to learn about the antique tractor hobby is to find someone near you who also enjoys the hobby yet has more experience than you, someone who can teach you the ropes and the tricks of the trade. In return, perhaps you can help this person with a restoration as the second set of hands we all need at times. This arrangement also lends itself well to the sharing of expensive rental tools, mutual travel arrangements to and from shows, and so on. Of course, as your skills, confidence, and abilities grow, you should serve as a mentor. That way you pass on what you have learned, but equally important (and like your mentor probably found out and as any school teacher will tell you), sometimes the best way to learn is to teach.

Clubs and Organizations

There are literally thousands of organizations across the world that can help you learn more about antique tractors and provide shows and structured activities to exhibit and acquire your restoration projects. Every one of these organizations is willing to accept new members willing

to pull their share of the weight, and your volunteer hours are a small price to pay for the opportunity to learn and grow in the antique farm equipment restoration pastime. Many of these clubs often have a particular orientation, though, so be sure to visit many times before making your mind up about which club to join. Some specialize in antique tractor pulls and putting on pulling competitions; some specialize in simply staging a community show every year; and some specialize in providing a network of support that members rely on for their restorations and learning. Regardless of the orientation, be sure it matches what you need and want from a local club. The appendices have more information about clubs and organizations to join.

Governmental Agencies and Industry Groups

Across many parts of the world, there are governmental agencies whose charge it is to support and help the farmer. In the United States, the predominant governmental system for hands-on help for the American farmer is an agricultural extension system. While there isn't much assistance for the collector and restorer, there is much help for the person trying to farm with antique equipment. This program varies quite a bit from state to state, but is very similar in the most important ways. Extension services provide day-to-day assistance to farmers in all areas of farming, and they can be a wealth of information. While I have never worked for the Agricultural

Extension service, I did work with a county government department that worked closely with them, and I have seen many cases where their advice made the difference between a farmer making and losing money, or between a saved crop and a ruined crop. Be sure to get to know your agriculture extension agent if you plan to use your antique tractor for any type of farming or land-renovation program.

Other organizations include industry trade groups, magazines, university and standards committees, and professional associations. The Early Days Gas Engine and Tractor Association (referred to as EDGETA) has branches across the country that hold and sponsor many shows, pulls, and clinics. Universities such as the University at California at Davis have significant farm equipment collections. The American Society of Agricultural Engineers is the first professional organization you should look to for in-depth technical material. Its published papers serve as standards for tractor design. Its standards publications are a must-read for anyone trying to decide how to fabricate a three-point retro fit to their tractor, find out PTO specification, or any number of other important bits of information. These organizations all play a part in providing the leadership and information antique tractor owners need to use and restore their equipment.

Another organization you should be familiar with is the University of Nebraska's Tractor Test Laboratory. This laboratory is more responsible for providing the world with actual, usable specifications about almost all the tractors made, past and present, than the manufacturers themselves. The tests were started by the state of Nebraska in the early 1900s as a way to protect the farmers of Nebraska from fraudulent claims of tractor manufacturers. In the process, the laboratory has compiled a huge history of the capabilities of farm tractors throughout the industry's history. This information is invaluable when selecting an antique tractor to buy, when transporting a tractor on a trailer (the tests recorded the weight of the tractor), and when trying to decide what types of work your current tractor may be capable of.

Information on the Internet

Increasingly, the Internet has become an important source of information for antique tractor enthusiasts. The act of using, repairing, and restoring antique tractors is data and information intensive: How was a particular part used? What year is my tractor? Where can I get manuals for my tractor? Where can I find new or used parts? All this information could be easily organized, stored, and retrieved by computers, and in fact is done so regularly. Many sites exist to help you find this information. I own Antique Tractor Internet Services, a company that specializes in providing a communication forum on the Internet about antique tractors. I also provide the computer resources for antique tractor–related businesses, collectors, and professionals to share their information and services with Internet users. By far the most widely used service of my company is the mailing list service. This service is free and provides a way for folks interested in antique tractors to ask questions to find out more about their tractors and restoration projects. Visit my site at http://www.atis.net. There is a wealth of information (all free), plus I provide a list of other sites that can help you if my site doesn't contain what you need.

Parts and Restoration Service Sources

Here is a list of all the sources of parts and services. This does not imply that I have done business with, nor personally vouch for, all of these companies. In fact, quite the opposite is true. I have not done business with nor know of most of these companies; however, all of these firms came recommended to me over the years, mostly through my Web site. That would imply that there is at least one happy customer for each of these names. I have included comments when they have been provided to me by the company or the individual recommending the company. Many thanks to Jim Poole, a wonderful friend in old iron who was responsible for putting together most of this list.

Salvage yards comprise most of the listings, many of which deal in both new and used parts. More important, many of these businesses are small sideline businesses for the owners, who have other more pressing priorities such as a day job or farm work. A mixture of persistence and patience will go a long way in helping you establish contact with these companies. Before I itemize the list of sources, I will give a couple of general recommendations:

▶ Dealers for major manufacturers should not be overlooked. John Deere is still a great source of parts for antique tractors, as is Ford/New Holland, Case (IH and Farmall, as well as Case), and AgCo.

▶ National auto parts stores, in particular, NAPA stores, are very well stocked for all kinds of industrial and agricultural engines, carry an excellent line of paints (Martin Senour), and maintain an efficient network of warehouses that minimize waiting for parts.

▶ Local farm equipment stores very frequently have some parts and services for antique tractors. Of particular mention is the brand of tractor parts named TISCO. TISCO is sold through these local stores, and they distribute a good selection of antique tractor parts.

▶ Swap meets are held across the country throughout the year, either alone or in conjunction with antique tractor shows. These are great sources for parts, particularly hard-to-find parts, books, manuals, and memorabilia.

▶ Classified advertisements in periodicals and Web sites are another good source for parts and services.

Where possible, I have included all contact information, though on many entries some piece of information or other is missing.

Internet Services

Antique Tractor Internet Services
3160 MacBrandon Lane
Pfafftown, NC 27040
(910) 924-6109
sales@atis.net
http://www.atis.net
Specializes in providing Internet services for antique tractor businesses, collectors, and clubs.

Magazines and Periodicals

9N-2N-8N-NAA Newsletter
P.O. Box 275
E. Corinth, VT 05040-0275
http://www.n-news.com
Jason P. Rinaldi, publisher

Antique Power
P.O. Box 562
Yellow Springs, OH 45387
(800) 767-5828
http://www.antiquepower.com
Patrick Ertl, publisher/editor
An all-brands tractor magazine, with free classifieds to subscribers.

The Belt Pulley
P.O. Box 58
Jefferson, WI 53549
http://www.beltpulley.com
920-674-9732
An all-brands tractor magazine.

Cockshutt Quarterly
International Cockshutt Club
http://www.cockshutt.com
info@cockshutt.com
419-522-1164
For Cockshutt, Co-op, Blackhawk, and Gambles Farmcrest enthusiasts.

Engineers and Engines
P.O. Box 10
Bethlehem, MD 21609-0010
410-673-2414
http://www.eandemagazine.com
stant@threshermen.org
All brands, with emphasis on steam engines.

Gas Engine Magazine
1506 SW 42nd St.
Topeka, KS 66609-1265
http://www.gasenginemagazine.com
For enthusiasts and collectors of all types of gas engines.

Green Magazine
2652 Davey Road
Bee, NE 68314
402-643-6269
http://www.greenmagazine.com
For John Deere enthusiasts.

MM Corresponder
3693 M Ave
Vail, IA 51465
712-677-2491
http://www.minneapolismolinecollectors.org/magazine/corresponder.asp
mmcorresponder@win-4-u.net
For Minneapolis-Moline enthusiasts.

Old Abe's News
J. I. Case Collectors' Association Inc. Newsletter
P.O. Box 638
Beecher, IL 60401
http://jicasecollector.com
For Case enthusiasts.

The Old Allis News
10925 Love Road
Bellevue, MI 49021
269-763-9770
allisnews@aol.com
For Allis-Chalmers enthusiasts.

Hart-Parr Oliver Collector
C/O Becky Losey
11326 N. Parma Rd.
Springport, MI 49284
becky@hartparroliver.org
http://www.hartparroliver.org/magazine.html
For Oliver enthusiasts.

The Prairie Gold Rush
17390 S. St. Rd. 58
Seymour, IN 47274
812-342-3608
http://www.prairiegoldrush.com
prairiegold@bremc.net
For Minneapolis-Moline enthusiasts.

Red Power Magazine
P.O. Box 245
Ida Grove, IA 51445
712-364-2131
http://www.redpowermagazine.com
For all IH (Farmall) enthusiasts.

The Rumely Newsletter
P.O. Box 12
Moline, IL 61265
309-764-7653
For Rumely enthusiasts.

Two-Cylinder Club
P.O. Box 430
Grundy Center, IA 50638-0430
888-782-2582
http://www.two-cylinder.com
For John Deere enthusiasts; annual membership includes magazine.

Manuals and Books

American Society of Agricultural and Biological Engineers
2950 Niles Road
St. Joseph, MI 49085
269-429-3033
ASABE is the organization that creates and supports agricultural standards. The organization sells copies of the standards and has several books about antique tractors.

Binder Books
P.O. Box 230269
Tigard, OR 97281-0269
(503) 684-2024
inquiry@binderbooks.com
http://www.binderbooks.com
This organization specializes in providing manuals for all kinds of International Harvester products, though it has selections of other manuals for other brands.

John Deere Technical Information Book Store
http://techpubs.deere.com/deere/Default.aspx
This is where you buy manuals from John Deere. You can also buy them at your local John Deere dealer.

Jensales Tractor Manuals
200 Main Street
Manchester, MN 56007-5000
(800) 443-0625
http://www.jensales.com
A large selection of manuals and farm toys.

NOS, New, Reproduction, and Salvage Parts

2 Cylinder Plus Salvage
322 Marlin Prairie Dr.
Conway, MO 65632
(417) 589-2634
http://2cylplus.com
rpm@2cylplus.com
John Deere, IH, and Oliver parts

ABC Company
P.O. Box 216
Letcher, SD 57359
(800) 843-3721

Abilene Machine
P.O. Box 129
Abilene, KS 67410-0129
(800) 255-0337
and
2797 130th Drive
Belmond, IA 50421
800-866-1504
http://www.abilenemachine.com

Ag Tractor Supply
9301 Breagan Road
Lincoln, NE 68526
800-944-2898
http://www.agtractorsupply.com

Agri-services
13899 North Road
Alden, NY 14004
(716) 937-6618
http://www.wiringharnesses.com
Wiring harnesses.

Albert Lea Tractor Parts
77847 209th Street
Albert Lea, MN 56007
800-338-6972
http://www.altractorparts.com

Alexander's Tractor Parts
P.O. Box 28
Winnsborro, TX 75494
(800) 231-6876
http://www.alexanderstractorparts.com

American Radiator
204 N. Oregon Ave.
Pasco, WA 99301
800-826-6649
and
1500 Yakima Valley Hwy
Sunnyside, WA 98944
888-837-6403
http://www.americanradiator.com

Anderson Tractor Supply
20968 Township Road 51
Bluffton, OH 45817
800-603-8141
http://www.andersontractorinc.com

Antique Auto Battery Manufacturing
Company
2320 Old Mill Road
Hudson, OH 44236
(800) 426-7580
http://www.antiqueautobattery.com

Antique Gauges Inc.
12287 Old Skipton Road
Cordova, MD 21625
(410) 822-4963
John Deere-authorized source of
replacement gauges. They carry other gauges
as well.

Arnold's
701 State Hwy 55 E.
Kimball, MN 55353
(612) 398-3800
http://www.arnoldsinc.com
There are four other Arnold's dealerships
located in Willmar, Glencoe, St. Martin,
and Mankato, Minnesota.

Austin Farm Salvage
Route 4, Box 291
Butler, MO 64730
660-679-4080
http://www.austinfarmsalvage.com

Austin Ignition Company
56 North Union Street
Akron, OH 44304
330-253-8132

Ed Axthelm
5071 Ashley Road
Cardington, OH 43315
(419) 864-4959

B&M Tractor Parts Inc.
101 Sloan St.
Taylor, TX 76574
(800) 356-7155
http://www.bmtractorparts.com

Baker Abilene Machine
P.O. Box 88
Bishopville, SC 29010
800-543-2451
http://www.bakerabilenemachine.com

Balcom Implement
1029 West Lake Avenue
Fairmont, MN 56031
(800) 658-2309
http://www.balcomauction.com
MM and others.

Bannon Tractor Parts
3400 Fort Worth Drive, Suite A
Denton, TX 76205
940-566-0091

Bates Corporation
12351 Elm Road
Bourbon, IN 46504
800-248-2955
http://www.batescorp.com

Biewer's Tractor Salvage
16242 140th Avenue S.
Barnesville, MN 56514
(218) 493-4696
http://www.salvagetractors.com

Bob Logan Tractor Co.
P.O. Box 216
Franklin Grove, IL 31031
(815) 456-2222

Bozeman Machinery
3409 Idalou Road
Lubbock, TX 79403
(800) 766-2076
http://www.bozemanmachinery.com

The Brillman Company
2328 Pepper Rd.
Mt. Jackson, VA 22842-2445
888-274-5562
http://brillman.com
Specializes in secondary ignition components
(spark plugs, spark plug wires, etc).

Burgh Implement
657 Perry Highway
Harmony, PA 16037
(412) 452-6880
http://www.burghtractorparts.com

Chartrand Equipment Co.
9353 State Route 3
Red Bud, IL 62278
and
6760 State Route 3
Ellis Grove, IL 62241
888-770-8801
http://www.chartrandequip.com
Construction equipment.

Colfax Tractor Parts
10447 Field Avenue
Colfax, IA 50054
(800) 284-3001
http://www.colfaxtractorparts.com

Cox's Used Parts
202 North 6th St.
Charleston, IL 61920
217-345-2775

Cross Creek Tractor Company
4315 U.S. Hwy 278 East
Cullman, AL 35055
(800) 462-7335
http://www.crosscreektractor.com

D & S Salvage
280 Dewitt Dr.
Sikeston, MO 63801
573-471-6220

Dennis Polk Equipment
72435 SR 15
New Paris, IN 46553
800-795-3501
http://www.dennispolk.com

Dennler Supply
3120 Genesee Julietta Rd.
Juliaetta, ID 83535
(208) 276-3771

Detwiler Tractor Parts
110 S. Pacific
Spencer, WI 54479
715-659-4252
http://www.detwilertractor.com

Draper Tractor Parts, Inc.
1951 Draper-Brown Rd.
Garfield, WA 99130
800-967-8185
http://www.drapertractor.com

Patrick Edwards Ltd.
Langley Farm, Langley Lane
Clanfield, Bampton
Oxfordshire OX18 2RZ
United Kingdom
44-0-1367-810259
http://www.patrickedwardsmachinery.co.uk
Mainly Ferguson and Fordson.

Farmers Tractor Parts
1120 5th Ave N
Ellendale, ND 58436
701-349-3536

Farmland Tractor Supply
32427 Old Hwy 34
Tangent, OR 97389
877-928-1646
http://www.farmlandtractor.com/mm5/
merchant.mvc

Fergiland
The Gateway
Highfields
Cockshut Lane, Melbourne
Derby DE73 8DG
United Kingdom
44-0-1332-865809
http://www.fergiland.com
Ferguson, especially T20s.

Florin Tractor Parts
8345 Florin Road
Sacramento, CA 95828
800-223-9916
http://www.florintractor.com

Gap Tractor
11103 FM 219
Clifton, TX 76634
and
P.O. Box 97
Cranfills Gap, TX 76637
800-972-7078
http://www.gaptractor.com

Gary's Implement Inc./Harimon Equipment
P.O. Box 235
Bridgeport, NE 69336
877-605-5970
http://www.garysimplement.com

General Gear
733 Desert Wind Road
Boise, ID 83716
(208) 342-8911
http://www.tractorparts.com
Crawlers and antique construction and
industrial equipment. This is a great source
for crawler information. John Parks, the
owner, is extremely knowledgeable

Grainger Industrial Supply
888-361-8649
http://www.grainger.com

H & R Construction
20 Milburn St.
Buffalo, NY 14212
(800) 333-0650
http://www.hrparts.com
There are three other locations in Riverside,
California; Ocala, Florida; and Edmonton,
Alberta.

Hamilton Equipment
567 South Reading Road
Ephrata, PA 17522-0478
717-733-7951
http://www.haminc.com
There are other Hamilton Equipment dealer
locations in Delaware, Maryland, New
Jersey, Ohio, Virginia, and West Virginia.

J. P. Tractor Salvage
1347 Madison 426
Fredericktown, MO 63645
(573) 783-7055
http://www.jptractorsalvage.com
Good source for International Harvester parts.

Johnson Diesel
1114 142nd Ave.
Wayland, MI 49348-9751
800-423-8216

Junction Tractor Parts
2425 Johnstown-Utica Road NW
Utica, OH 43080
740-892-2889

Knoxland
6 Warner Road
Warner, NH 03278
603-746-5260
http://www.knoxland.com
MF dealer, well known for parts.

Lakeside Farm Implements Inc.
N6891 County Road B
Montello, WI 53949-8133
608-296-2045

Larry Romance & Son, Inc.
543 West Main Street
Arcade, NY 14009
585-492-3810
http://www.newholland.com/dealers/
LarryRomanceandSon
cromance@localnet.com

Larry's Tractor
1515 W. 13th Street
Tipton, IA 52772
563-886-2469

Manning Tractor Co.
305 E Boyce St.
Manning, SC 29102
803-435-8807

Maplewood Implement Sales Co.
19355 State Route 47
Maplewood, OH 45340
937-492-3436
Combines, planters, hay tools, implements.

Marshall Machinery Inc.
348 Bethel School Road
Honesdale, PA 18431
570-729-7117
http://www.marshall-machinery.com
Specializes in A-C; has IH, JD, Oliver, and
others.

McLeods Tractor Parts Ltd.
5410 Sherin Road
Edmonton, AB
T5A 0A7
Canada
780-476-1234

Meridian Equipment
5946 Guide Meridian
Bellingham, WA 98226
888-333-0892
http://www.meridianeq.com

A-C, Farmall, JD, Case.
Meyers Tractor Salvage
39012 128th St.
Aberdeen, SD 57401
605-225-0185
http://www.meyerstractor.com

Mid-South Salvage Inc.
P.O. Box 545
Decatur, AL 35602
256-353-5661
http://www.midsouthsalvage.net

Nash Equipment Company
Route 26
Colebrook, NH 03576
603-237-8857
http://www.nashequipment.com

Oakley Combine Sales
4201 Highway 96
Oxford, NC 27565
919-693-4367

The Old Tractor Company
7460 E. Hwy 86
Franktown, CO 80116
303-663-5246
http://www.theoldtractorcompany.com
Greg Stephens is a John Deere dealer and is
very knowledgeable and helpful. He regularly
writes a column for Green Magazine.

The Old 20 Parts Company
Cavendish Bridge
Shardlow, Derby DE72 2HL
United Kingdom
44-0-1332-792698
http://www.old20.com
Most U.K. Tractors.

Olson Power & Equipment
38560 14th Ave.
North Branch, MN 55056
651-674-4494
http://www.olsonpower.com

Otto Gas Engine Works
2167 Blue Ball Rd
Elkton, MD 21921
(410) 398-7340
http://www.pistonrings.net

Pelican Tractor
5629 South 6th
Klamath Falls, OR 97601
800-528-2744
http://www.pelicantractor.com

Pete's Tractor Salvage
2163 15th Ave. NE
Anamoose, ND 58710-9677
800-541-7383
http://www.petestractor.com
A very large salvage yard with great selection.

Restoration Supply Company
96 Mendon Street
Hopedale, MA 01747
800-809-9156
http://www.tractorpart.com

Rice Equipment
20 North Sheridan Road
Clarion, PA 16214-1216
814-226-9200
http://www.riceequipmentinc.com

Rice Tractor Salvage
6160 Melon Rd.
Sutherlin, VA 24594
434-753-2114

Ridenour's
413 West State Street
Trenton, OH 45067
(513) 988-0586
http://my.ohio.voyager.net/~dlride/
JD L/LA/LI specialist; also makes
reproduction sheet metal.

Rippee's Used Tractor Parts
5566 Hillsboro Road
Farmington, MO 63640-7228
573-431-5939

Rock Valley Tractor Parts
1004 10th Avenue
Rock Valley, IA 51247
800-831-8543
http://www.rvtractorparts.com

Rogue Valley Farm Equipment Repair
6002 Crater Lake Ave
Central Point, OR 97502
(541) 826-8500
Magnetos; highly recommended for any
tractor work.

Russells Tractor Parts
3710 E. Willow St.
Scottsboro, AL 35768
800-248-8883
http://www.russelltractor.com

Scholten's Equipment, Inc.
8223 Guide Meridian
Lynden, WA 98264
800-433-5480
and
9534 Green Rd.
Burlington, WA 98233
800-726-8081
http://www.scholtensequipment.com

Schwanke Tractor & Combine
3310 S. Highway 71
Willmar, MN 56201
800-537-5582
http://www.schwanketractor.com

Sharman Machinery
New England Highway
Glen Innes
NSW 2370
Australia
02 6732 3079
o.sharman@nsw.chariot.net.au
If he doesn't have it, he can get it.

Shelton's Family Farm
P.O. Box 66
Hartwood, VA 22471
540-752-2720
http://www.sheltonfarm.com
A-C specialist

Shoup Manufacturing Company
3 Stuart Drive
Kankakee, IL 60901
800-627-6137
http://www.shoupparts.com

Shuck Implement Company
1924 E. 1450 Road
Lawrence, KS 66044
785-843-8093

Skillings Tractor
4100 Dialton Road
Springfield, OH 45502-9631
937-964-1486

Smith's Implement Company Inc.
E1140 State Road 170
Downing, WI 54734
715-643-5911
800-826-2806

Southeast Tractor Parts
14720 Hwy 151
Jefferson, SC 29718
888-658-7171

Southern Counties Tractor Spares
137 Almodington Lane
Earnley, Chichester
West Sussex PO20 7JR
United Kingdom
44-0-1243-512109
http://www.southerntractorspares.co.uk
Ferguson and Fordson.

Spallinger Combine Parts
2325 County Road 28
Bluffton, OH 45817
800-758-8333
http://www.spallingercombine.com

Staben Equipment
821 N. Rice Ave.
Oxnard, CA 93030
800-404-2103
http://www.thetractorstore.com

Stamm Equipment Company
3450 12th Street
Wayland, MI 49348
269-792-6204

Steiner Tractor Parts
1660 S. M-13
Lennon, MI 48449
800-234-3280
http://www.steinertractor.com
Replacement parts; no used or OEM parts.

Stevens Tractor
Highway 71
Coushatta, LA 71019
800-333-9143
http://www.stevenstractor.com

Storey Tractor
11239 Karo Rd.
Bell Buckle, TN 37020
800-241-5670
storeytractor@bellsouth.net

Strojny Implement
1122 Highway 153
Mosinee, WI 54455
715-693-4515
Specializes in Ford Ns.

Stromp's Dump
Spalding, NE 68665
308-497-2211

Stuckey Brothers Farm Supply
3727 Hemingway Hwy
Hemingway, SC 29554
800-365-4232
http://www.stuckeybrosinc.com

Surplus Tractor
3215 W. Main
P.O. Box 2125
Fargo, ND 58107
800-859-2045

Timber Tractor Parts
9624 S. Hanna City Glasford Road
Glasford, IL 61533
309-389-5855
Ford specialist.

TMS Stoller Tractor
8310 Blough Road
Sterling, OH 44276
330-669-3676
International Harvester

Tony's Tractor Parts
P.O. Box 40
Lidgerwood, ND 58053
701-538-4300

The Tractor Barn
6154 West Highway 60
Brookline, MO 65619
800-383-3678
http://www.tractorbarn.net

Tractor Implement Supply Company
(TISCO)
Woods Equipment Company
2340 West County Road C
Roseville, MN 55113
800-338-0145
http://www.tiscoparts.com
http://www.woodsequipment.com

Tractor Parts Unlimited
8 Goshen Road
Bozrah, CT 06334
860-889-0051

The Tractor Place
1920 Mize Road
Knightdale, NC 27545
919-266-5846
http://www.thetractorplace.com

W.C. Littleton & Son Inc.
100 West 10th Street
Laurel, DE 19956
800-842-7445
http://www.wclittleton.com

Wall Lake Used Parts
201 S. Center Street
Wall Lake, IA 51466
800-522-1909

Washington Country Tractor
1889 Hwy 290 East
Brenham, TX 77834
979-836-4591
and
1310 Spur 515
Navasota, TX 77868
936-825-7577
http://www.wctractor.com
A-C, Farmall, Ford, Oliver.

Wayne's Used Parts
3060 County Road 1400N
El Paso, IL 61738
800-626-8915
http://www.waynesusedparts.com

Welters Farm Supply
14307 Lawrence 2190
Verona, MO 65769
(417) 498-6496

Wenger's of Myerstown
814 S. College St.
Myerstown, PA 17067
800-451-5240
http://www.wengers.com

Willard Equipment
2782 S R 99
Willard, OH 44890
419-933-6791

Wilson Farms
20552 Old Mansfield Road
Fredricktown, OH 43019
740-694-5071
Primarily a salvage yard, but some new parts.

Worthington Ag Parts
27170 US Highway 59
Worthington, MN 56187
(800) 533-5304
http://www.worthingtonagparts.com
There are other Worthington Ag Parts
stores located in Indiana, Iowa, Michigan,
Missouri, North Carolina, South Dakota,
Wisconsin, Canada, and Australia.

Worley Auction Service and Farm Salvage
7117 W. State Road 60
Campbellsburg, IN 47108
812-883-4313

Yesterday's Tractor Co.
751 Commerce Loop
PT Business Park
Port Townsend, WA 98368
800-853-2651
http://www.antiquetractorstore.com
Specializes in parts for pre-1970 tractors.

Zimmerman Oliver-Cletrac
1450 Diamond Station Road
Ephrata, PA 17522
717-738-2573
http://www.olivercletrac.com
Great source of Cletrac parts.

Restoration Services

2 Cylinder Diesel Shop
731 Farm Valley Road
Conway, MO 65632
417-589-3843
Specializes in two-cylinder John Deere
diesel engines.

Branson Enterprises
7722 Elm Avenue
Machesney Park, IL 61115
815-633-4262
http://shopping.safeport.com/cgi-bin/webplus.
cgi?script=/webpshop/store.wml&storeid=6
Carburetor and magneto restoration.

Burrey Carburetor Service
5026 Maples Road
Fort Wayne, IN 46816
800-447-6347
http://www.burreycarb.com

Denny's Carburetor Shop
8620 N. Casstown-Fletcher Road
Fletcher, OH 45326
937-368-2304
http://www.dennyscarbshop.com
They also sell solid state ignitions.

Fireball Heat Treating Company Inc.
34 John Williams Street
Attleboro, MA 02703
508-222-2617
This firm can perform almost any metallurgical
process involving heat treating, stressing, case
hardening, and so on. John Thomson, one of the
owners, is very knowledgeable, kind, and helpful.

Magneeders
8215 County Road 118
Carthage, MO 64836
417-358-7863
http://www.magneeders.com
Magneto restoration services.

Minn-Kota Repair, Inc.
38893 Hwy 12
Ortonville, MN 56278
320-839-3940
Steering wheel recovering.

Paper Pulleys Inc.
810 Woodland Street
Columbia, TN 38401
931-388-9099
http://www.paperpulleys.com
They restore paper-core belt pulleys.

RAM Remanufacturing and Distributing Inc.
N. 604 Freya
Spokane, WA 99202
800-233-6934
http://www.ramengine.com
The Vintage Engine Machine Works
Division can complete any type of engine
machining you need for your antique tractor.

Robert's Carburetor Repair
404 East 5th St.
Spencer, IA 51301
712-262-5311
http://www.robertscarbrepair.com
Also sells videos on how to do it yourself.

Restoration Supplies and Tools

A-1 Leather Cup and Gasket Company
2103 Brennan Ave.
Fort Worth, TX 76106
817-626-9664
Very good source of seals and gaskets.
Custom-cut felt seals.

Jorde's Decals
935 9th Avenue NE
Rochester, MN 55906
507-288-5483
http://www.jordedecals.com
Very highly regarded source for John Deere
decals.

K & K Antique Tractors
5995 N. 100 W.
Shelbyville, IN 46176
317-398-9883
http://www.kkantiquetractors.com
John Deere-authorized source of vinyl-cut
decals.

Lubbock Gasket and Supply
402 19th Street
Lubbock, TX 79408
800-527-2064
http://www.lubbockgasket.com
Custom and standard gaskets for antique
tractors.

POR 15 Inc.
P.O. Box 1235
Morristown, NJ 07962
800-726-0459
http://www.por15.com
Their gas tank sealing kit is especially
effective.

TP Tools & Equipment
P.O. Box 649
Canfield, OH 44406
800-321-9260
http://www.tptools.com
A very nice selection of quality sandblasting
products.

Tires

M. E. Miller Tire
17386 State Highway 2
Wauseon, OH 43567
800-621-1955
http://www.millertire.com
Also sells memorabilia, tire putty, and
parade lugs.

Tucker's Tire
844 S. Main Street
Dyersburg, TN 38024
888-248-7146
http://www.tuckertire.com
This place is well-liked and well-received
among antique tractor enthusiasts.

Glossary

Ampere: A measure of electrical current.
This measures how many electrons are
passing through a given point in the circuit.
Analogous to flow rate in plumbing systems.

ASABE: American Society of Agricultural
and Biological Engineers.

Babbit: A lead and tin alloy that creates,
in the presence of engine oil, an excellent
bearing surface for parts to turn and move in.
In older antique tractors, this material must
be poured into place, while newer tractors
have easily inserted and removed bearing
shells lined with a babbit type of material.

Ballast: Weight added to the tractor to
improve traction.

BDC: Bottom dead center. The lowest point
of travel of the piston.

Belt pulley: A small drum-shaped pulley
on a tractor used to drive wide, flat belts
that in turn drive other equipment such as
threshers, mills, choppers, and others.

Bendix: A name brand of electric starter engagement systems that engages the ring gear of a flywheel until the engine starts and then retracts.

Bull gear: The large gears in the tractor's drivetrain that provide the majority of the reduction in power delivery. See Reduction.

Bushing: A metal sleeve (bronze, soft steel, or brass) that in the presence of grease serves as a bearing surface for a moving or rotating part.

Camshaft: The shaft inside an engine responsible for activating valves at the proper time and interval. It is driven by the crankshaft in most antique tractor engines.

Channel frame: This structural design adds pressed or cast steel frames that the engine mounts to. These frames also serve to connect the front of the tractor to the transmission, allowing the engine to become a nonintegral part of the structural system of the tractor.

Choke: A carburetor's choke restricts the air intake passage for the purpose of increasing the fuel-to-air mixture. This aids starting the engine when it is cold.

Compression ratio: The ratio of the volume of air in an engine cylinder when the piston is at bottom dead center, and the volume of that same air when the piston has traveled to top dead center. Compression ratios of antique tractor engines are typically in the range of 6:1 to 9:1.

Crankshaft: The shaft inside the engine that harnesses the straight-line movement of the piston as a rotational force.

Curtains: See Shutters.

Differential: A device found in tractor transmissions that redirects the power delivered by the engine to the individual rear drive axles in a way that applies the power differentially, based on operating conditions. This is the device that allows the outside wheel, during a turn, to rotate more often than the inside wheel.

Disc harrow: Implement with one or two rows of offset discs. Used in seedbed preparation.

Displacement: The amount of volume that an engine has inside the cylinders and combustion chambers. This is expressed in cubic inches (ci).

Distillate: Renderings of crude oil that were not diesel, gasoline, or the pure fuels. Typically, this fuel was in between gas and diesel in weight and vaporization characteristics.

Draft sensing hitch: An implement hitching system that senses the amount of load on the tractor and compensates the implement height or presentation to compensate. The Ferguson hitching system found on Fergusons, Ford 8Ns, and most modern tractors are examples.

Drift: A brass or soft steel tool used to help drive and set other, harder steel parts. The soft metal prevents damage to the harder parts being installed or removed.

Final drives: The collective name of the rear-wheel-drive components between the differential and the rear drive wheels. This includes the large reduction gears called bull gears.

Flywheel: Any object designed to store energy through the rotation of a large mass. These are present in engines and help to smooth out power delivery.

GCWR: Gross combined weight rating. The total combined weight that a vehicle can safely pull or tow. This includes passengers and items in the tow vehicle itself.

Glow plug: A small heating device used to warm up a diesel combustion chamber or manifold before starting that helps to make starting easier.

GPS: Global positioning system. A device that calculates, with the aid of satellites, your exact position on earth. Used in farming to locate soil test sites, monitor yields, and to apply soil additives.

GVWR: Gross vehicle weight rating. The total amount of weight a vehicle can safely handle. This includes passengers and driver, baggage, and the tongue weight of any trailer, but not the trailer itself.

Governor: A device found on tractor engines, both gas and diesel, that is designed to regulate engine rpm. This is accomplished through automatic changes in throttle settings as needed in response to field conditions.

Horsepower: A unit of energy. Specifically, one horsepower is the amount of energy required to lift 33,000 pounds 1 foot over a 1-minute time period.

Impulse coupling: A device inside a magneto that helps to start a tractor engine by automatically creating stronger spark during the slow engine speeds of starting.

Injector: The device in a diesel engine that delivers fuel to the combustion chamber.

Injector pump: The fuel pump on a diesel engine that delivers fuel to the injectors at very high pressures.

Integral frame: A structural design of many antique tractors in which the engine block casting serves as a structural member of the tractor. This eliminates extra structural pieces to connect the front axle and engine to.

L head engine: The name of an engine design that places the valves in the engine block instead of in the cylinder head. This is a synonym for a flathead engine.

LPG: Liquefied petroleum gas. This fuel as a power source for tractors began to see some use after World War II, but fell out of favor by the end of the 1950s.

Live power: A generic term used to describe any device or system on the tractor that accepts power inputs directly from the engine itself. This is opposite of non-live power, in which the power source for the system is other than the engine, such as the transmission, rear axles, and so on.

Live PTO: See Live power and PTO.

Live hydraulics: See Live power.

Lugs: That portion of rear steel wheels that are bolted to the outer circumference and provide traction.

Magneto: A device that uses the principle of electromagnetic interactions between magnets and wires to create a spark in the combustion cylinder of an engine at the proper time.

Manifold: The part on any engine that directs the intake air and exhaust in and out of the engine.

MIG: Metal inert gas welding, now more properly known as GMAW, or gas metal arc welding. This process allows very fine control of heat applied to the weld, which allows for very thin pieces of metal, such as sheet metal, to be welded together.

Multifuel: An adjective applied to any tractor that can operate on any number of fuels. These tractors usually start on gasoline, though.

NFE: Narrow front end. A style of tractor that has two front wheels mounted very closely together, giving it the appearance of a tricycle. Occasionally, the front end has just one wheel.

NOS: New old stock. This refers to parts that were manufactured during the same time period the tractor was manufactured, but have never been installed on a tractor. Some restorers believe these parts are the most appropriate when restoring a tractor.

Oil bath air cleaner: This air-cleaner design percolates the engine's incoming air through a reservoir of oil, forcing oil up onto baffle plates and a steel oil filter element. The dirt in the incoming air is then trapped in the oil.

Over-running clutch: A device that attaches to non-live PTO shafts to prevent the inertial force of rotating implements from driving the tractor forward, even after the clutch has been pressed in and brakes applied.

Pedestal: The large iron casting on tractors—usually narrow front-end tractors—that supports the front wheels and serves as a frame and support for the steering assembly.

Pony motor: The small gas motor that is used as a starter in large diesel-engine tractors.

PTO: Power take-off. An engine- or transmission-driven shaft that protrudes from a tractor, usually from the rear, that can be used to power auxiliary devices.

Reduction: The act of reducing a power input's rpm for the purpose of increasing the amount of delivered torque. A small gear turning a large gear is an example of a reduction gear set.

Refurbishing: The act of sprucing up a tractor, insofar as the tractor is cleaned, painted, and obvious faults and problems are addressed. A tractor that has been refurbished will still clearly look and possibly act its age and is not considered to be restored.

Remanufacturing: The act of renovating a tractor so completely and extensively that the untrained eye would not be able to distinguish between the antique and the tractor when it was new.

Remote hydraulics: The capability of tractors to operate auxiliary devices driven by hydraulic rams or motors.

Restoration: The act of renovating a tractor to the point the tractor looks and behaves as if it were new.

Resistor: An electrical device that limits the flow of electricity. Twelve-volt ignition coils often require that a resister be placed in the circuit to help condition the spark during operation. Six-volt systems, which most antique tractors have, do not require this.

ROPS: Roll-over protection systems. These systems are designed to protect the operator should the tractor overturn. The system usually comprises a roll-bar or cab, and a seat belt.

Row crop: A style of tractor whose wheels are designed to straddle crops grown in rows and whose stances were high enough to clear these same crops.

Solenoid: A remote-control switch that operates through the application of electric current. The current creates a magnetic field that causes the main switch to be thrown. These are found on electric starters.

Shutters: Devices placed in front of the radiator that, when closed, help to heat up the engine. This is used in conjunction with a fuel other than gasoline.

TDC: Top dead center. The point where the piston is at the very top of its travel.

Tedder: Implement that tosses hay, which expedites the drying process.

Thermosiphon: The principle that air or water, in a closed system, will flow and move if heat is added to it. Older tractors do not have water pumps, relying rather on thermosiphoning to move and turn over cooling water.

Three-point hitch: A hitch that attaches implements to a tractor using three different linkages, hence the name. It became popularized by tractors that had the Ferguson draft control system.

Tie rod: Those rods within the steering linkage of a tractor used to connect the pitman arm drag linkage and the axle spindles.

Tires, loaded: Tires whose tubes are filled with water (actually an antifreeze and water solution) for the express purpose of adding ballast to the tractor.

Torque: A measure of rotational force, calculated in foot-pounds.

Tractor fuel: The name given to distillate fuel after its characteristics became standard and agreed upon.

Unstyled tractors: This refers specifically to John Deere Letter Series tractors before the styling influences of Henry Dreyfuss. Generally, this term is used to refer to any tractor whose design and workmanship minimizes styling and cosmetic influences.

Volt: A measure of the electrical force.

WFE: Wide front end. Any antique tractor with a front axle that widely spaces the front wheels.

Wheatland: Any style of tractor that was designed to work large wheat fields. These tractors were typically large, had nonadjustable front and rear wheels, no hitch except the drawbar, and full fenders.

Wrist pin: Same as a piston pin; it joins the piston to its connecting rod.

Zerk fitting: A small valve allowing grease, under pressure, to be admitted while preventing contaminants from entering. It prevents grease from escaping back through it.

Index